Introduction to 4G Mobile Communications

For a listing of recent titles in the
Artech House Moblie Communications Series,
turn to the back of this book.

Introduction to 4G Mobile Communications

Juha Korhonen

ARTECH HOUSE

BOSTON | LONDON
artechhouse.com

Library of Congress Cataloging-in-Publication Data
A catalog record for this book is available from the U.S. Library of Congress.

British Library Cataloguing in Publication Data
A catalog record for this book is available from the British Library.

ISBN-13: 978-1-60807-699-4

Cover design by Vicki Kane

© 2014 Artech House

All rights reserved. Printed and bound in the United States of America. No part of this book may be reproduced or utilized in any form or by any means, electronic or mechanical, including photocopying, recording, or by any information storage and retrieval system, without permission in writing from the publisher.

All terms mentioned in this book that are known to be trademarks or service marks have been appropriately capitalized. Artech House cannot attest to the accuracy of this information. Use of a term in this book should not be regarded as affecting the validity of any trademark or service mark.

10 9 8 7 6 5 4 3 2 1

Contents

Acknowledgments	xi
1 Introduction	**1**
2 History of Mobile Telecommunications	**5**
2.1 First Generation	5
2.1.1 Introduction	5
2.1.2 AMPS	6
2.1.3 TACS	7
2.1.4 NMT	7
2.1.5 C-NETZ	8
2.1.6 Radiocom2000	8
2.1.7 Japanese Systems	8
2.1.8 Summary	9
2.2 Second Generation	9
2.2.1 Introduction	9
2.2.2 Global System for Mobile Communication (GSM)	10
2.2.3 Digital Advanced Mobile Phone System (D-AMPS)	13
2.2.4 Code Division Multiple Access (IS-95 CDMA)	14
2.2.5 Personal Digital Cellular (PDC)	16
2.2.6 Integrated Dispatch Enhanced Network (iDEN)	16
2.2.7 Terrestrial Trunked Radio (TETRA)	17
2.3 Third Generation	17
2.3.1 Introduction	17
2.3.2 3GPP	18
2.3.3 3GPP2	22
2.4 Conclusions	26
References	28
3 Overview of a Modern 4G Telecommunications System	**29**
3.1 Introduction	29
3.2 LTE-A System Architecture	29
3.3 LTE RAN	30
3.4 OFDM Air Interface	31

3.5	Evolved Packet Core	33
3.6	LTE Requirements	33
3.7	LTE-Advanced	40
3.8	LTE-A in Release 11	44
References		45

4 OFDMA   47

4.1	Introduction	47
4.2	OFDM Principles	51
4.3	LTE Uplink—SC-FDMA	56
4.4	Summary of OFDMA	57
References		58

5 Air Interface: Physical Layer   59

5.1	Introduction	59
5.2	Physical Layer—General	60
5.3	Physical-Layer Processing	62
	5.3.1 CRC Insertion	63
	5.3.2 Channel Coding	64
	5.3.3 Physical-Layer Hybrid-ARQ Processing	68
	5.3.4 Rate Matching and Channel Interleaving	69
	5.3.5 Scrambling	70
	5.3.6 Modulation	71
	5.3.7 Layer Mapping and Precoding	75
	5.3.8 Mapping to Assigned Resources and Antenna Ports	76
5.4	Physical Channels in LTE-A	76
	5.4.1 Downlink Physical Channels	77
	5.4.2 Uplink Physical Channels	80
5.5	Physical Layer Signals	81
	5.5.1 Downlink Signals	81
	5.5.2 Uplink Signals	85
5.6	Summary	86
References		86

6 Air Interface: Protocol Stack   89

6.1	Introduction	89
6.2	Medium Access Control (MAC)	90
	6.2.1 MAC—General	90
	6.2.2 Transport Channels	96
6.3	Radio Link Control (RLC)	99

	6.3.1	RLC—General	99
	6.3.2	RLC Functions	101
	6.3.3	Logical Channels	102
6.4	Packet Data Convergence Protocol (PDCP)		105
6.5	Radio Resource Control (RRC)		108
	6.5.1	Introduction	108
	6.5.2	UE States	108
	6.5.3	RRC Functions	110
References			123

7 Radio Access Network — 125

7.1	Introduction		125
7.2	E-UTRAN Architecture		125
7.3	eNodeB		127
	7.3.1	eNodeB Introduction	127
	7.3.2	eNodeB Functionality	128
7.4	Home eNodeB		134
	7.4.1	Introduction	134
	7.4.2	Closed Subscriber Group (CSG)	136
	7.4.3	HeNB Mobility	137
	7.4.4	HeNB Gateway	137
	7.4.5	Traffic Offloading	137
7.5	Relay Node		142
References			145

8 Core Network: Evolved Packet Core — 147

8.1	Introduction		147
8.2	Architecture		148
	8.2.1	Mobility Management Entity	148
	8.2.2	Serving GW	150
	8.2.3	Packet Data Network Gateway	150
	8.2.4	Home Subscriber Server	151
	8.2.5	Evolved Serving Mobile Location Center	151
	8.2.6	Gateway Mobile Location Center	155
	8.2.7	Policy Control and Charging Rules Function	155
8.3	EPC Interfaces and Protocols		155
	8.3.1	EPC Control Plane	155
	8.3.2	EPC User Plane	160
	8.3.3	Summary of EPC Interfaces	162
References			166

| 9 | Procedures | 169 |

9.1 Introduction 169
9.2 Cell Search 169
9.3 Random Access 171
9.4 Tracking Area Update 174
9.5 Initial Context Setup 179
9.6 Handover (X2 interface) 182
9.7 CSG Inbound HO 186
9.8 S1 Release Procedure 189
9.9 Dedicated Bearer Activation 191
References 193

| 10 | Specifications | 195 |

10.1 Introduction 195
10.2 Internal Structure 195
   10.2.1 TSG RAN 197
   10.2.2 TSG SA 199
   10.2.3 TSG CT 200
   10.2.4 TSG GERAN 201
   10.2.5 Mobile Competence Center (MCC) 202
10.3 Standardization Process 202
   10.3.1 Introduction 202
   10.3.2 Work Items 202
   10.3.3 Version Numbering 204
   10.3.4 Releases 205
   10.3.5 Development Cycle 209
10.4 Specification Numbering 210
10.5 Backwards Compatibility 211
10.6 E-UTRAN Specifications 212
References 214

| 11 | LTE-A Features | 215 |

11.1 Energy Saving 215
11.2 MIMO 219
   11.2.1 MIMO Overview 219
   11.2.2 Downlink MIMO 221
   11.2.3 Uplink MIMO 224
11.3 Relays 224
11.4 Carrier Aggregation 226

11.5	Enhanced Intercell Interference Coordination	229
11.6	Evolved Multimedia Broadcast Multicast Service	233
11.7	Self-Organizing Networks	235
11.7.1	Automatic Neighbour Relations (ANR)	235
11.7.2	Mobililty Load Balancing (MLB)	235
11.7.3	Mobility Robustness Optimization (MRO)	235
11.7.4	Coverage and Capacity Optimization	236
11.7.5	RACH Optimization Function	236
11.7.6	Coordination Between Various SON Functions	236
11.8	Coordinated Multipoint Transmission/Reception	236
11.8.1	Downlink CoMP	237
11.8.2	Uplink CoMP	238
11.9	LTE and Voice	238
References		241

12   Future Developments   243

12.1	Introduction	243
12.2	Evolving LTE-A	244
12.2.1	Proximity Services	244
12.2.2	Machine-Type Communications	248
12.2.3	Mobile Relays	251
12.2.4	Heterogeneous Networks	255
12.3	The Fifth Generation	257
12.3.1	Introduction	257
12.3.2	New Air Interface	258
12.3.3	Bandwidth	259
12.3.4	Network Architecture	261
12.3.5	Multiple Antennas	262
12.3.6	Support for M2M Communications	263
12.3.7	Other Improvements	264
12.4	METIS Project	265
References		268

Appendix: LTE and LTE-A Specifications	269
About the Author	277
Index	279

Acknowledgments

I would like to thank my wife, Akiko, for running the household and taking care of our baby while I was up in our attic room writing this book. For many months she usually met me only when I climbed down, looking for a meal in our kitchen. And baby Koyuki, who from now on can spend much more time playing with daddy.

I would also like to thank my colleagues at ETSI and especially the people on the 3GPP project team. They have been very supportive of my book-writing project. Moreover my manager, Mr. John Meredith, is a walking data bank on all matters 4G, which is a very useful feature in one's manager.

Within Artech House, I would especially like to thank Ms. Aileen Storry. She has had to endure a lot with me and my slipping deadlines.

Chapter 1

Introduction

Long-term evolution (LTE) was originally only an internal 3GPP name for a program to enhance the capabilities of 3G radio access networks. This program was initiated already in 2004. The name was never meant to become the official name for the new 4G mobile communications system. However, the name stuck, and its use has spread from the engineering community to general public. LTE has become synonymous with 4G.

At the same time with LTE development, 3GPP was also working with another evolutionary project known as system architecture evolution (SAE), which produced the specifications for the evolved packet core (EPC) network. SAE or EPC have not become as well-known buzzwords as LTE. However, a LTE access network could not work properly without an EPC core network supporting it, and therefore EPC is also discussed in this book extensively.

This book is about LTE-advanced (LTE-A), an enhanced version of LTE. LTE-A fulfills the requirements set by the International Telecommunications Union (ITU) for a 4G mobile communications system. Currently there is no plan to replace this name with a catchier marketing name, so it seems that the new LTE-A system will be there for good. However, it seems that quite often the general public simply uses the term LTE without any distinction between LTE and LTE-A.

The first LTE specifications were published in December 2008, in a specifications release known as 3GPP Release 8. The first LTE-A specifications followed in 3GPP Release 10, which was published in mid-2011. At the time of this writing in late 2013, 3GPP is working with Release 12, to be published in late 2014. 3GPP standardization is a continuous process; new features are added in every release and older features are improved. Therefore it is fair to say that 3GPP specifications are never ready—they are constantly evolving.

This book provides an introduction to the world of LTE-A. It is only an introduction—it does not explain every detail in LTE-A, and in any case that would be impossible. 3GPP standards contain hundreds of specifications, which jointly have probably tens of thousands of pages in them. This also means that it would be very difficult to study a given feature just by reading the specifications. Even finding the correct specifications that handle a given feature is a major achievement for a new LTE student. Moreover, 3GPP standards are written by a group of specialist engineers to other specialist engineers. The contents can be very cryptic for nonspecialists.

Typically a standard does not explain, it just sets the rules and requirements. The user of those standards has to have deep understanding of the issues on hand already; otherwise, the text in standards is incomprehensible.

This book explains the LTE basics in plain English, oftentimes with a few examples, pointing out how various issues are interconnected, and then refers to further sources of information for those who want to learn more about a specific subject.

History of mobile telecommunications is discussed in Chapter 2. It is important to know the past because it helps us to understand the present. How have we ended up with the systems we have today? This chapter discusses the development of mobile telecommunications generation by generation. It shows how we have arrived in a situation where we have only one serious competitor in the fourth generation, whereas in the first generation many countries had (at least) one national system of their own.

Chapter 3 presents the overview of a modern 4G telecommunications system; that is, an LTE-A system. First, the LTE system is compared to 3G-UMTS: what are the changes and why are they needed. When the LTE project was launched, a list of high-level requirements was presented for the system to be designed. These requirements are listed in Chapter 3, and then they are discussed one by one: can the LTE system fulfill the requirements and how does it do it? The same treatment is given for the LTE-A: its requirements are listed, and then the performance of the new system is analyzed against them. In the end of the chapter the most important new LTE-A features are presented briefly. Note that Chapter 11 will analyze some of these features in depth.

LTE and LTE-A have adopted a new air interface technology called ODFMA. In Chapter 4, the new air interface is compared to the old one used in 3G. Why was WCDMA not suitable for 4G anymore, and why can OFDMA supposedly fulfill the requirements better? OFDMA basics are discussed, as are the reasons LTE air interface employs different uplink and downlink technologies.

Chapter 5 concentrates on the air interface physical layer. This chapter is a continuation of Chapter 4. Whereas Chapter 4 discussed OFDMA technology, Chapter 5 concentrates on other aspects of the physical layer. The downlink data path through the physical layer is described, and the physical layer functionality along that path is discussed in detail. Also, physical channels and physical layer signals are explained.

Chapter 6 is a sizeable chapter because it presents the air interface protocol stack. MAC, RLC, PDCP, and RRC layers are handled here. Each of them is discussed in its own section. The most important functions from each layer are explained. Transport and logical channels are also introduced and discussed.

Chapter 7 is about the radio access network. In LTE the radio access network is called E-UTRAN (as in evolved UTRAN). In fact, the

E-UTRAN is very different from the UTRAN in 3G. The E-UTRAN has no radio network controllers. The only node type it has is called eNodeBs, which stands for evolved NodeB. The eNodeB is a combined base station and base station controller. This has resulted in some fundamental changes in how the radio access network operates and especially in how it is managed. There is, for example, a new direct logical interface, X2, between the base stations. The E-UTRAN also includes home eNodeBs and relay nodes, which are discussed in this chapter.

Chapter 8 handles the core network, or evolved packet code (EPC) in 3GPP jargon. Even though its name seems to suggest that it is somehow just an enhanced version of the UMTS core network, the EPC is very different from its predecessor. Partly this stems from the fact that the UMTS had both circuit-switched and packet-switched domains, whereas the EPC has only a packed-switched domain. This chapter is divided in two parts. In the first part, EPC architecture is presented and its entities and their functions are discussed one by one. The second part concentrates on EPC interfaces and the protocol stacks in them.

Chapter 9 includes procedure descriptions. The most important procedures in the LTE/EPC system are described in detail, with the help of diagrams and figures. In earlier chapters, the system functionality was discussed layer by layer or entity by entity. This may have left some questions on how everything fits together. This chapter shows how various entities cooperate to provide services.

Chapter 10 discusses 3GPP standardization. For those readers who want to gain in-depth knowledge on a particular topic, the 3GPP specifications provide the ultimate source of information. However, it is not easy to find what you want to find from those publications. Therefore this chapter first explains the structure of 3GPP, its working practices, its release regime, how specifications are modified, and what is inside each specifications series. When the reader is equipped with all this information, he may have a chance to find something from the specifications.

LTE-A is an evolution of LTE, with a set of new features and improved performance. Chapter 11 discusses a selection of new LTE-A features. The list is not complete—it cannot be since there are so many new features in LTE-A. However, the author believes he has selected the most important and interesting new features for this chapter.

Finally, Chapter 12 contains some crystal-ball gazing. We try to have a peek into the future and guess what is going to happen in mobile telecommunications in the latter part of this decade.

CHAPTER 2

History of Mobile Telecommunications

2.1 First Generation

2.1.1 Introduction

Mobile telecommunications actually has quite a long history already. The first generation cellular systems were launched around 1980, but it must be noted that there were already mobile phones in existence long before that. The first mobile phone systems were launched soon after the Second World War, first in the United States and thereafter in various European countries. But these were radio telephones, not cellular systems (see Figure 2.1). They did not support concepts such as handovers or roaming. The call was only maintained as long as the phone remained within the coverage area of the transmitting tower. The number of simultaneous calls the system could handle was very small because each call required a separate frequency channel. Mobile phones were large, and typically they were installed in cars. The weight and size of these devices did not enable handheld usage. In some sources, these systems are known as generation zero.

The first generation cellular systems were analog systems (for voice channels, that is; control channels were already digital in most systems). Each frequency carrier could carry only one call. Obviously, as a result, the capacity of such systems was small. However, it was much better than in radio telephone systems because the cellular concept enabled frequency reuse. The cellular approach also provided an easy way to increase capacity where it was needed. In the countryside, the cells could be larger, providing wide coverage but lower system capacity, whereas in cities, the cells would be smaller, greater in number, and provide more capacity. Being cellular systems (see Figure 2.2), these networks also supported handovers. The user could move from one base station area to another, and the ongoing call was maintained.

First generation networks were typically only countrywide. The technology was distinct in different countries, and thus roaming to other networks was not supported except in some special cases such as in NMT

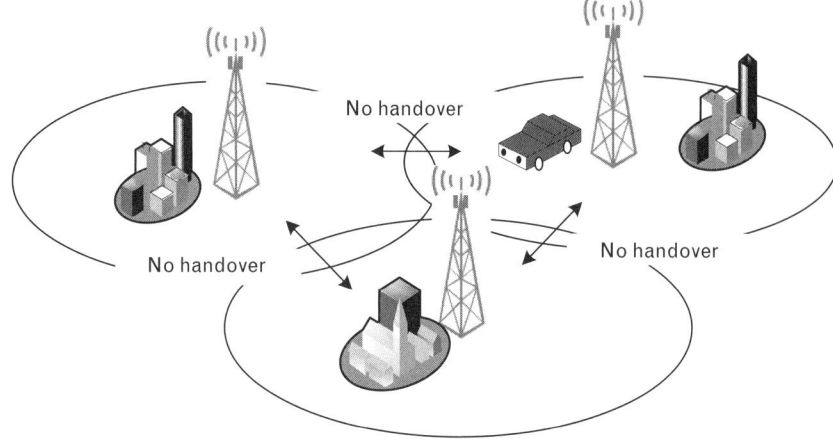

FIGURE 2.1
Radio telephony system.

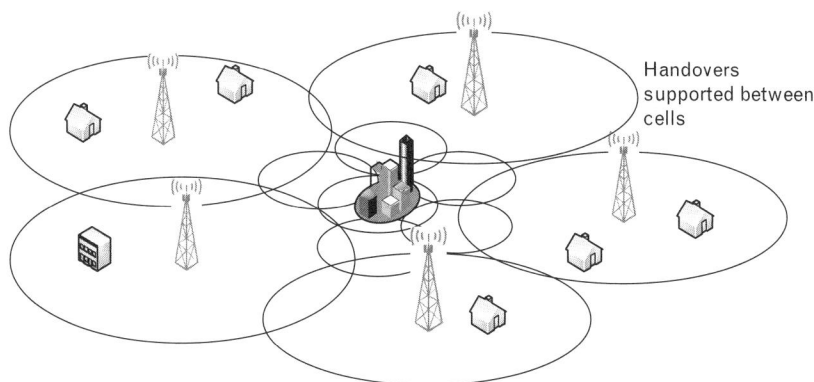

FIGURE 2.2
Cellular system.

systems in Northern Europe. The list of first generation technologies and the countries that adopted them are given in Table 2.1 (note that this list is not exhaustive).

2.1.2 AMPS

Advanced mobile phone service (AMPS) was a US-developed (by Bell Labs) standard, operating in the 850-MHz band. It was also widely used in several countries in Asia and South America. The first commercial AMPS network was launched in 1983 in the US. The AMPS system employs full duplex channels with a 45-MHz channel separation. Individual channels are 30 kHz each. This is an all-analog system (i.e., both control and traffic channels are analog).

Narrowband advanced mobile phone service (NAMPS) was an analog cellular system that was introduced by Motorola in late 1991 and thereafter deployed in several countries. NAMPS used narrow 10-kHz bandwidth

TABLE 2.1 FIRST GENERATION TECHNOLOGIES AND DEPLOYMENT COUNTRIES

SYSTEM	COUNTRIES
AMPS	Argentina, Australia, Bangladesh, Brazil, Brunei, Burma, Cambodia, Canada, China, Georgia, Guam, Hong Kong, Indonesia, Israel, Kazakhstan, Kyrgyzstan, Malaysia, Mexico, Mongolia, Nauru, New Zealand, Pakistan, Papua New Guinea, Philippines, Russia, Singapore, South Korea, Spain, Sri Lanka, Tajikistan, Taiwan, Thailand, Turkmenistan, United States, Vietnam, Western Samoa
NMT-450	Argentina, Australia, Belarus, Belgium, Cambodia, Croatia, Czech Republic, Denmark, Estonia, Faroe Islands, Finland, France, Germany, Hungary, Iceland, Indonesia, Italy, Latvia, Lithuania, Malaysia, Moldova, Netherlands, Norway, Poland, Romania, Russia, Slovakia, Slovenia, Spain, Sweden, Thailand, Turkey, Ukraine
NMT-900	Cambodia, Cyprus, Denmark, Faroe Islands, Finland, France, Greenland, Netherlands, Norway, Serbia, Sweden, Switzerland, Thailand
TACS/ETACS	Austria, Azerbaijan, Bahrain, China, Hong Kong, Ireland, Italy, Japan (JTACS), Kuwait, Macao, Malaysia, Malta, Philippines, Singapore, Spain, Sri Lanka, United Arab Emirates, United Kingdom
C-Netz	Germany, Portugal, South Africa
Radiocom 2000	France
RTMI	Italy
MCS-LI, MCS-L2	Japan

radio channels, which is one-third of the AMPS channel bandwidth. As a result, the system capacity was increased.

2.1.3 TACS

Total access communication system (TACS) was a UK-based standard, although its roots lie in AMPS. TACS operated on the 900-MHz band, and it was used mainly in the UK and in Ireland. The first TACS network was launched in 1985. TACS system was deployed on 25-kHz radio channels, which were narrower than the 30-kHz channels used in AMPS. Extended TACS (ETACS) was an extended version of TACS, providing more frequency channels.

2.1.4 NMT

Nordic mobile telephone (NMT) was different from other 1G systems in that it was specified jointly by Nordic telecommunication authorities, and thus it supported roaming between different operators from the beginning. And because the specifications were open, it enabled several companies to manufacture NMT hardware and compete in the marketplace. NMT came in two variants, NMT-450 and NMT-900. The numbers indicate the operating frequency in megahertz. The first NMT-450 networks were launched in 1981, and NMT-900 followed in 1986.

NMT became very popular and was adopted in a large number of countries, even outside the Nordic region. The reasons for this were that the system was technically quite advanced, and it was based on an open standard that enabled lower cost for equipments. NMT-450, with its lower frequency band and higher maximum transmit power, could provide large coverage (up to 40-km radius), whereas NMT-900 was more suited for areas that required higher system capacity. In both systems, the channel bandwidth was 25 kHz.

2.1.5 C-NETZ

C-Netz was a German-developed 1G standard, replacing the earlier B-Netz radio telephony network. C-Netz operated on the 450-MHz band, and its channel bandwidth was 20 kHz. It did not support roaming, and indeed there would have been little use for that functionality since the only other countries adopting C-Netz were Portugal and South Africa. The first C-Netz system became operational in Germany in 1985.

2.1.6 Radiocom2000

Radiocom2000 was a French system developed by Matra. It was launched in 1986 and operated on the 400-MHz band. Two years later, an NMT-based network was also launched in France.

2.1.7 Japanese Systems

Japan deserves a special mention because in mobile communications it has always been among early adopters. NTT DoCoMo launched its MCS-L1 (mobile cellular system) network in 1979 [1]. This system was based on AMPS technology, which was then yet to be launched in the United States. It operated on the 800-MHz band. In 1988 an improved MCS-L2 system was launched, operating on the same frequency channels. In some sources MCS-L2 is also known as Hi-Cap. MCS-L2 improvements include halving the radio channel bandwidth, from 25 kHz to 12.5 kHz, and diversity reception in mobile handsets. Because NTT DoCoMo's mobile services became very successful, competing operators started similar services using Japanese Total Access Communication System (JTACS) technology in 1989. JTACS was based on TACS (and therefore also AMPS). JTACS was swiftly improved in narrowband TACS (NTACS) in 1991. NTACS, like MCS-L2 before it, employed 12.5-kHz channels, therefore increasing the system capacity considerably.

2.1.8 Summary

Many of these systems continued to operate until after 2000 but were then gradually closed, and the frequencies were reused for 2G/3G systems. The main problem with these systems was that they were not designed for mass markets. The system designers did not anticipate that mobile phone penetration could raise to 100 percent or over as it is today. Rather, it was thought that these devices would only be used by special groups who travel often and need to communicate a lot, such as marketing people and senior managers. When more and more customers started to use the networks, the capacity simply ran out. There were also other problems, such as security. In many of these systems the security issue was not very well designed. Calls could be eavesdropped, and in many cases it was also possible to steal the identity of the phone, clone it, and thereafter make calls for free.

2.2 Second Generation

2.2.1 Introduction

The second generation was the first one to employ all-digital transmission technology, both for signaling and traffic. This greatly enhanced the capacity of systems and quality of services. In first generation systems, one call required one frequency channel (in full-duplex systems this was actually two frequencies; one for the uplink and one for the downlink), but in second generation systems one frequency channel was divided between several users by means of time-division or code-division techniques, which resulted in increased system capacity. There were also many other improvements in 2G that increased the system capacity even further. It is good to note that the development of a major mobile communications standard is a continuous process. New releases of specifications are published regularly, with improvements to networks, services, and mobile handsets. Especially in case of GSM, this continuous development process has lasted more than 20 years by now. The latest Enhanced Data rates for Global Evolution (EDGE)-enhanced GSM systems are completely different from those GSM networks that were launched in early 1990s. However, at some point it may become obvious that the basic platform has become too old and cannot easily be improved further, in which case a new system will be designed, with a new generation "tag." Many 2G technologies have already been abandoned, and there is less and less development work being done with EDGE. It is foreseen that soon, maybe in 1–2 years, GSM and EDGE will no longer be improved further. Operational GSM/EDGE networks will be here for many years to come though.

There were several attempts to draft standards for digital mobile communication systems. Not all proposals survived. There are four main 2G systems that have been in widespread use:

- GSM (originally Groupe Special Mobile, later Global System for Mobile communications);
- Digital advanced mobile phone system (D-AMPS);
- Code division multiple access (CDMA);
- Personal digital cellular (PDC).

Moreover, integrated dispatch enchanced network (iDEN) networks have been launched in several countries, although its subscriber numbers are clearly smaller than with the systems listed here.

2.2.2 Global System for Mobile Communication (GSM)

GSM [2] was designed to be the pan-European standard, but later it became very popular and has spread all over the world, with the most notable exception being Japan. The original GSM was using the 900-MHz band because that was the band allocated for the pan-European system. However, later on more bands were allocated for GSM: 850 and 1900 MHz in Americas, and 1800 MHz in Europe, and existing bands were extended. GSM was also specified for 450 MHz (GSM-450), to replace NMT-450 networks, but this system has not seen commercial deployments. The first GSM network was launched in 1991 in Finland.

GSM is a time division multiple access (TDMA) system. In TDMA each frequency carrier is divided into timeslots that are allocated for different users dynamically (see Figure 2.3). In basic GSM there are eight timeslots in a frame (i.e., eight users can have simultaneous calls on the same frequency). In its simplest form, a GSM phone will transmit on only one timeslot and

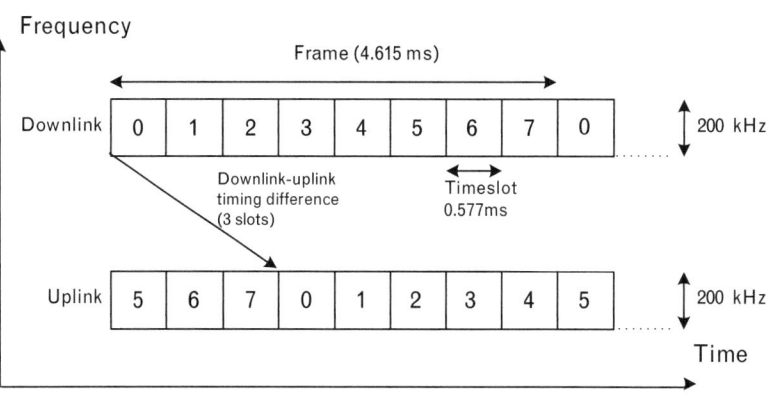

FIGURE 2.3
GSM frame structure.

receive on one (but it does not have to do these things simultaneously since the uplink is delayed by three timeslots when compared to the downlink). That leaves six timeslots when the phone can measure its radio environment or stay idle and save power. The GSM channel bandwidth is 200 kHz.

2.2.2.1 Circuit Switched Data (CSD)

As pointed out earlier, GSM has undergone continuous development and new improvements are still being introduced into the standard. The most notable enhancements of GSM were the faster data services. In the beginning GSM supported mainly voice, and its data services were primitive. The first GSM data service, circuit switched data (CSD), provided only a 9.6-Kbps connection. This was enhanced with the high speed circuit switched data (HSCSD) service to 14.4 Kbps per timeslot, and HSCSD was also able to use up to four timeslots per frame, giving the maximum data rate of 57.6 Kbps. The problem with these services was that the timeslot(s) employed were allocated for a user for the whole duration of the call. For many applications the data transmission requirements are not static but highly variable. Therefore circuit switched data services are likely to waste capacity and are therefore expensive to use. However, they do offer low latency, which is important for some applications, and the data rate is guaranteed. Also, for operators the HSCSD update was a rather easy step to take since it did not require large investments in hardware.

2.2.2.2 General Packet Radio Service (GPRS)

The next data service to be introduced was GPRS. This packet switched service was fundamentally different from the earlier circuit switched service. In GPRS the timeslots were allocated for the data service only when there was data to send or receive. Thus, the service could adapt to highly variable data transmission needs. However, this flexibility did not come cheap for network operators. To support GPRS, the core network required a major upgrade with a set of new components and modifications to existing ones. However, GPRS was seen as a necessary step for operators to take in their development path. Without GPRS, a 2G operator would not have survived for long because it could not have provided data services effectively.

The maximum data rate in early GPRS systems was 40 Kbps, but in later releases this was increased to 171 Kbps. The maximum rate will require all timeslots in a frame to be assigned to a single user, but this is still not as bad a situation as a four-slot HSCSD, because GPRS does not reserve these slots permanently. They are only used when there is data to be sent. If there is no data to be sent, those slots can be allocated to some other users. This re-allocation can be done on a frame-by-frame basis. The drawback of this arrangement is that there will be more latency because mobile devices must first get a permission to transmit from the data scheduler in the network. Moreover, this latency does not have a guaranteed upper limit, though the operator can implement a quality of service (QoS) scheme in order to try to limit the delays. Therefore, the basic GPRS is not suitable for interactive services that require guaranteed low latency, but it is better with noninteractive services such as email, streaming audio/video, and ftp. It is also suited for web browsing since even though browsing is an interactive process, the delay requirements are not very strict.

The first operational GPRS networks were launched in 2001.

2.2.2.3 Enhanced Data rates for Global Evolution (EDGE)

The next improvement to GSM was EDGE. The idea behind EDGE is a new modulation scheme called eight-phase shift keying (8PSK). It increases the data rates of standard GSM by up to threefold. This modulation scheme does not replace the old Gaussian minimum shift keying (GMSK) modulation but rather coexists with it. 8PSK can only be used effectively in good radio conditions over short distances and/or higher transmission power, and thus GMSK is needed for weaker radio conditions. The first version of EDGE provided maximum data rate of 384 Kbps, but this has then increased to 1.3 Mbps (downlink) and 653 Kbps (uplink) with EDGE Evolution. To achieve this kind of performance, EDGE Evolution employs techniques such as:

- Higher order modulation: the basic EDGE employ 8PSK in addition to GMSK, but EDGE evolution has also 16QAM and 32QAM (QAM = Quadrature Amplitude Modulation) in its toolbox. Whereas GMSK carries 1 bit per symbol, and 8PSK 3 bits per symbol, in 16QAM and 32QAM these numbers are 4 and 5, respectively. However, 16QAM and 32QAM can only be used in very good radio conditions.

- Receiver diversity: as is well known, the receiver diversity improves the quality of the reception.

- Parallel data channels: EDGE evolution can make use of two parallel data channels.

- Simultaneous transmission and reception: standard GSM/EDGE handsets have only one combined transmitter/receiver chain in order to save costs. However, EDGE evolution proposes handsets with independent transmission/reception parts, so it would be possible to use up to eight timeslots per frame (i.e., the whole frame) for transmission and reception.

However, it is good to point out that even though EDGE evolution can potentially provide performance comparable to 3G, it has not been taken up very well by operators. Even though the actual evolved EDGE upgrade to networks is a relatively simple process for operators, the operators also want to see evolved EDGE–capable handsets in the market that can use those services before they commit to this upgrade. And handset manufacturers do not want to manufacture evolved EDGE handsets before there is a market for those handsets. In all, there is a feeling that GSM upgrades are coming to a close and evolved EDGE could provide only a temporary solution before the operator would need to upgrade its network to UMTS/LTE. Also, it is good to note that even though evolved EDGE looks good on paper performancewise, most of this gain is achieved by using more spectrum resources (i.e., parallel data channels and simultaneous Tx/Rx). Higher order modulation is an improvement that cannot always be used. In weak radio conditions, it may actually make things worse and result in reduced data throughput.

When it comes to the "generation game," then GPRS is regarded as 2.5G and EDGE is 2.75G.

2.2.3 Digital Advanced Mobile Phone System (D-AMPS)

D-AMPS is another popular 2G system that is known by several other names too: North American TDMA, North American digital cellular (NADC), US-TDMA, IS-54, IS-136, or simply TDMA. Therefore D-AMPS is a bit vague since it could refer to either IS-54 or IS-136, which are quite different systems, with IS-136 being the successor of IS-54.

IS-54 is a digital extension of the first generation AMPS system. It is backwards compatible with it in the sense that both systems use the same analog control channel structure and the same channel bandwidth (30 kHz). That is, IS-54 is only half digital in the sense that its control channels are analog. This provided a smooth path for AMPS operators to upgrade their systems to digital. IS-54 divided a traffic channel into three timeslots (full rate), or six timeslots (half rate).

However, IS-54 soon ran out of capacity, and an upgrade was needed. The result, IS-136, was an all-digital system (both control and voice channels), which was also incorporated the functionality of the IS-54 standard (see Figure 2.4). Therefore it was possible to use AMPS and IS-54–capable phones in an IS-136 network. IS-136 also introduced new features such as text messaging, CSD, sleep mode, picocells, and improved security. All IS-136 channel frames were divided into six timeslots.

The IS-54 system was operational in 1990, and IS-136 followed in 1994. The operating frequency for IS-54 was the same as for AMPS (i.e., 850 MHz). IS-136 could also operate on the 1900-MHz band. Note that

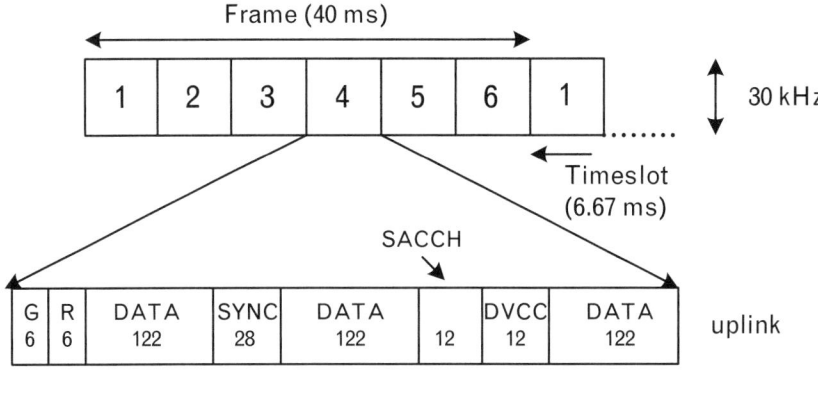

FIGURE 2.4
IS-54 and IS-136 frame structures..

since 1900 MHz was an IS-136 specific band, it did not need to support analog AMPS phones (i.e., the 1900 MHz band was all digital).

D-AMPS was also enhanced with a cellular digital packet data (CDPD) specification in 1993. The maximum data rate provided was 19.2 Kbps. Even though this feature was deployed in some D-AMPS networks in the United States, it did not receive much enthusiasm from consumers. It was, however, used to provide some special services such as machine-to-machine communications (M2M) and public safety services. The last CDPD networks were finally shut down in 2005.

2.2.4 Code Division Multiple Access (IS-95 CDMA)

Code division multiple access (CDMA) standard IS-95 was developed by Qualcomm in the late 1980s. IS-95 was different from other commercially deployed 2G systems in that IS-95 was based on CDMA technology, with the others being TDMA systems. Whereas in TDMA the radio channel is divided into timeslots that are then allocated to different users, in CDMA the radio channel is divided into code channels by means of different spreading codes (see Figure 2.5). That is, in CDMA the transmissions to each user are continuous.

FIGURE 2.5
The difference between TDMA and CDMA.

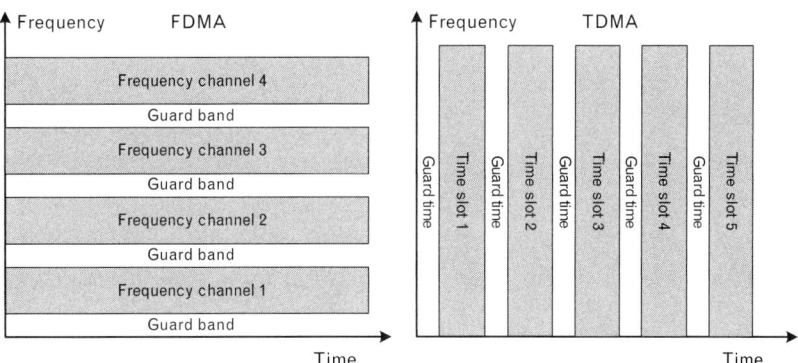
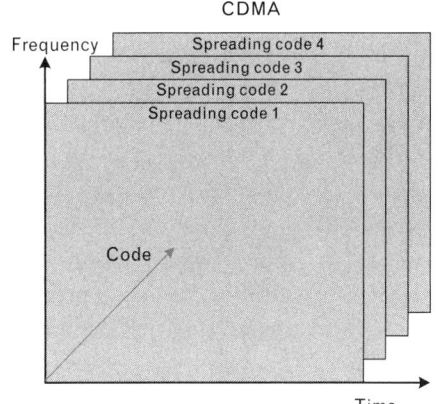

CDMA has also some other properties that make it different from other 2G technologies. The same frequency carrier may be used in adjacent cells, which means no frequency planning is needed (though a different planning exercise, namely code planning, is required). CDMA also enables soft handovers (i.e., at cell edges the mobile device can be connected to more than one base station simultaneously, improving the call quality). The IS-95 radio channel bandwidth is 1.25 MHz.

CDMA as a technology has become exceptionally popular since it is was also adopted by UMTS later (though not in the form that was initially promoted by Qualcomm).

IS-95 was also known as cdmaOne; with IS-95 being the standard name and cdmaOne the brand name given to it by the CDMA Development Group (CDG). IS-95 became popular in North America and some parts of Asia. However, it could not compete with GSM when it comes to European markets. The first commercial IS-95 network was launched in 1995 in Hong Kong. Thereafter IS-95 got two major revisions: IS-95A and IS-95B. Further revisions (e.g., 1xRTT) are regarded as 3G systems already.

2.2.5 Personal Digital Cellular (PDC)

PDC is the Japanese 2G standard. Originally it was known as Japanese Digital Cellular (JDC), but the name was changed to Personal Digital Cellular to make the system more attractive outside Japan. However, this renaming did not bring about the desired result, and the PDC standard was used only in Japan. Though it is good to point out that CDMA (IS-95) was also adopted by some operators in Japan. The air interface specification is known as RCR-27, and the system operated on two frequency bands: 800 MHz and 1500 MHz. PDC had both analog and digital modes; the analog mode was for backwards compatibility to older first generation networks in Japan. The PDC system employs TDMA technology. Its physical layer parameters are quite similar to D-AMPS, but its protocol stack resembles GSM. It uses 25-kHz carriers, and each frame is divided into three or six timeslots.

NTT DoCoMo was the first operator to launch PDC in 1993. It became an extremely popular service, with more than 80 million subscribers at its peak (in a country of about 127 million people). Also PDC evolved toward faster data connections. NTT DoCoMo developed a proprietary data service called i-mode. It employed a packet data network (PDC-P) behind the PDC radio interface. The popularity of mobile services in Japan also meant that Japan was very eager to design and adopt a 3G system as early as possible, before their 2G systems ran out of capacity. As a result, NTT DoCoMo became the leading promoter of WCDMA technology.

Once 3G networks became available in Japan, PDC subscriber numbers started to diminish, and the networks were closed one by one. The last PDC network was closed by NTT DoCoMo in 2012.

2.2.6 Integrated Dispatch Enhanced Network (iDEN)

Integrated dispatch enhanced network (iDEN) is a mobile telecommunication system that includes features from both cellular systems and trunked radio systems. iDEN was developed by Motorola, and it provides voice, data, and push-to-talk services. It was first deployed in 1996 by Nextel. The radio channel bandwidth is 25 kHz, and TDMA frames are divided into six time slots per frame. The operating frequencies are on 800-MHz and 900-MHz bands.

iDEN users are typically governmental organizations and large companies. The operators can be "normal" telecom operators or small private service providers. iDEN has a lots of users, especially in Central and Southern America. However, the number of iDEN users is already declining, and, for example, in the United States the old Nextel iDEN network was closed by its new owner Sprint Nextel in 2013.

2.2.7 Terrestrial Trunked Radio (TETRA)

TETRA's usage areas are similar to iDEN. TETRA has been designed for governmental agencies, military, emergency services and transport services to provide closed group communication services in addition to normal cellular services. The first version of TETRA was published in 1995 by European Telecommunications Standards Institute (ETSI), and the second version, TETRA 2, in 2005. In addition to normal cellular services, TETRA offers also push-to-talk services. Its operating frequencies are typically quite low, around 400 MHz, depending on a particular country, and thus it can operate on much larger cells than GSM/UMTS/LTE networks. The channel bandwidth in TETRA 1 is 25 kHz, and the TDMA frame is divided into four timeslots. The problem with TETRA has been that its data transmission capabilities were rather slow, especially if compared to "public" cellular networks. This issue will partly be solved by TETRA 2 standard once those networks are deployed. However, given the rapid development of cellular standards and capabilities, it is likely that TETRA will always lag behind other cellular technologies when it comes to data transmission capabilities.

TETRA is currently widely deployed (almost) all over the world. In the United States, the use of TETRA was not authorized before, but FCC allowed the use of TETRA on non–public safety bands in 2012. The reasoning for not allowing TETRA for public safety bands were interoperability problems and potential interference caused by TETRA.

2.3 Third Generation

2.3.1 Introduction

The extraordinary success of 2G mobile communication networks with ever increasing subscriber numbers made it obvious for mobile network operators that new capacity was needed, and it was needed quickly. The design work for third generation systems had already started about the same time as the first second generation systems were deployed.

In Japan the PDC network was quickly running out of capacity, and NTT DoCoMo wanted to push forward with 3G faster than other industry members. NTT DoCoMo's technology of choice was WCDMA. In Europe there were support for both TDMA- and CDMA-based solutions, but NTT DoCoMo's example probably tipped the scale in favour of CDMA (i.e., WCDMA). In the United States, there were two popular 2G systems, D-AMPS and IS-95. D-AMPS supporters formed a consortium called Universal Wireless Communication Consortium (UWCC) to promote a TDMA-based solution by enhancing the existing IS-136 standard. The new standard was known as UWC-136. Similarly, IS-95 supporters grouped together to promote an IS-95-based solution that was called CDMA2000.

Lots of research both in the industry and in the academia was carried out for various technologies. As seen, the promoters of the most widely deployed 2G technologies had different ideas on how to proceed. The International Telecommunication Union (ITU) might have preferred to select only one 3G technology, but from early on it was clear that this was not going to happen, for both technical and political reasons. Instead ITU drafted IMT-2000, the umbrella specification of all 3G systems.

In its November 1999 meeting Helsinki, the ITU accepted the following proposals as IMT-2000 compatible:

- IMT direct spread (IMT-DS; also known as UTRA FDD);
- IMT multicarrier (IMT-MC; also known as CDMA2000);
- IMT time code (IMT-TC; also known as UTRA TDD);
- IMT single carrier (IMT-SC; also known as UWC-136);
- IMT frequency time (IMT-FT; also known as DECT).

However, in the end only three of these proposals survived and were developed further under two consortia; 3GPP and 3GPP2. UTRA FDD and UTRA TDD were developed by the Third Generation Partnership Project (3GPP). This technology was commonly known also as wideband code division multiple access (WCDMA), or universal mobile telecommunications system (UMTS). CDMA2000 was progressed further by another industry consortium, 3GPP2. IMT-SC was originally supported by UWCC, but later this organization decided to adopt UTRA FDD as its 3G technology. UWCC disbanded itself and in 2002 a new organization, 3G Americas, was formed to promote WCDMA systems in the Americas. When it comes to the DECT-based 3G system, it never took off and that initiative was abandoned.

2.3.2 3GPP

2.3.2.1 Introduction

The 3GPP is an organization that developed specifications for a 3G system based on the UTRA radio interface and on the enhanced GSM core network. 3GPP is also responsible for the future GSM development work. Note that since then 3GPP has also taken over the 4G (LTE) standards development work. That is, currently 3GPP is doing standards development for three mobile communications generations; 2G, 3G, and 4G. The participants in 3GPP are mainly from the telecommunications industry: equipment vendors, network operators, regulators, academia, and other research establishments. The membership is arranged around organizational partners: the

Association of Radio Industries and Businesses (ARIB), Japan; the Alliance for Telecommunications Industry Solutions (ATIS), United States; China Communications Standards Association (CCSA); the European Telecommunications Standards Institute (ETSI); Telecommunications Technology Association (TTA), Korea; and Telecommunication Technology Committee (TTC), Japan. A company can apply 3GPP membership by first becoming a member of one of the organizational partners.

2.3.2.2 UMTS

The air interface technology developed by 3GPP is called wideband code division multiple access (WCDMA) [3]. This system is also known as the universal mobile telecommunications system (UMTS). It has two modes: frequency division duplex (FDD) and time division duplex (TDD). In the FDD mode, the uplink and downlink use separate frequency bands. In the first version of the standard the WCDMA carriers had a 5-MHz bandwidth, although in later versions also other bandwidths were specified. Each carrier is divided into 10-ms radio frames, and each frame further into 15 timeslots. The UTRAN chip rate is 3.84 Mcps. A chip is a code word used to modulate the information signal. Every second, 3.84 million chips are sent over the radio interface. However, the number of data bits transmitted during the same period by a user is much smaller. The ratio between the chip rate and the data bit rate is called the spreading factor. In principle, the spreading factor indicates how large a chunk of the common bandwidth resource the user has been allocated. For example, one carrier could accommodate at most 16 users, each having a channel with a spreading factor of 16. The spreading factors used in UTRAN can vary between 4 and 512. The sequence of chips used to modulate the data bits is called the spreading code. Each user is allocated a unique spreading code (or several codes).

The TDD mode differs from the FDD mode in that both the uplink and the downlink use the same frequency carrier. The 15 timeslots in a radio frame can be dynamically allocated between uplink and downlink directions; thus, the channel capacity of these links can be different. The chip rate of the normal TDD mode is 3.84 Mcps, but there also exists two other options, the 1.28 Mcps low chip rate (LCR) option and the 7.68 Mcps option. The 1.28 Mcps option was introduced in Release 4, and the 7.68 Mcps option was added to Release 7. The nominal carrier bandwidths for 1.28 Mcps, 3.84 Mcps, and 7.68 Mcps options are 1.6 MHz, 5 MHz, and 10 MHz, respectively. The TDD mode is actively promoted by Chinese companies, and China Mobile has been given licences to operate TDD mode networks in China, in both 3G and 4G technologies. The advantage of the TDD mode is that the timeslots in a frame can be allocated asymmetrically for data services where the expected bit rates in the uplink and downlink differ significantly. Also, because TDD does not require symmetric band allocations, it can be deployed on individual spectrum bands, which are easier to find than "paired" symmetric bands.

The UMTS core network (in its early form) is an enhanced GSM/GPRS core network—many elements of the GSM/GPRS network can be reused in 3G, which makes the migration from 2G to 3G easier for network operators. The radio access network (base stations and base station controllers) in UMTS is new though.

2.3.2.3 Releases

The standards work in 3GPP is organized around releases. A release is a self-contained set of standards (this set can be rather large—a 3GPP release typically contains hundreds of individual specifications). There is no exact schedule on when new releases will be published. At early stages of UMTS work there were new releases every year, but later on this time has grown to 18 months or so. More about 3GPP working practises and releases in Chapter 10.

When it comes to 3G, the first 3GPP release was known as Release 99 (though it was already the year 2000 before that release was published). Release 99 provided 64-Kbps circuit switched and 384-Kbps packet switched connections. After Release 99 the numbering of releases was changed as it became obvious that a year-based numbering would be difficult to maintain; not all years would have a new release. After Release 99 came Release 4 (originally called release 2000), which was released in 2001. This release was rather thin in content. The most important new features in Release 4 included:

- Transcoder free operation (TrFO);
- Tandem free operation (TFO);
- Virtual home environment;
- Full support of location services (LCS);
- The new narrowband, or low chip rate, TDD mode (1.28 Mcps);
- UTRA repeater;
- Robust header compression (ROHC).

Release 5 followed in 2002. This is a bigger release than Release 4. It contains several large and important enhancements to 3GPP systems. One could also quite justifiably claim that Release 5 provides the first true 3GPP system, as only Release 5's new features make it possible to provide genuine 3G-specific services with attributes that people are expecting from

3G. Therefore, it is slightly ironic that marketing people already like to call HSDPA 3.5G. The new features in Release 5 include the following:

- High-speed downlink packet access (HSDPA);
- Wideband AMR (WB-AMR) codec;
- IP-based multimedia services (IMS);
- Intradomain connection of RAN nodes to multiple CN nodes (Iu-Flex);
- Reliable end-to-end QoS for packet-switched domain.

HSDPA, especially, was a very important enhancement. It increased the maximum data rate in the downlink to 14.4 Mbps (with 16QAM modulation and code rate of 1). However, HSDPA is not so much about increasing the theoretical maximum data rate (which, in mobile telecommunications, is very theoretical anyway); it is more about increasing the practical throughput. The highly dynamic and adaptive nature of HSDPA means that HSDPA channels will be able to transmit the highest possible amount of data frame after frame even in poor and changing radio conditions.

After Release 5 it took a longer time before the next release was ready. Release 6 came out at the end of 2004. This is also a large release, containing features such as:

- High-speed uplink packet access (HSUPA);
- Multimedia broadcast multicast services (MBMS);
- Push to talk over cellular (PoC);
- Generic access network (GAN);
- Improvements to IMS.

The main features here are HSUPA and MBMS. HSUPA can provide 5.76 Mbps maximum data rate in the uplink in Release 6. MBMS is potentially a very important feature for operators, enabling them to offer broadcast services such as TV and radio. However, even though this release was out in 2004, there are no operational MBMS systems yet anywhere in 2013. The problem is probably in that MBMS consumes a large chunk of the network capacity, and in 3G the capacity is still in short supply. However, in LTE the corresponding feature (eMBMS) will be launched at least by Verizon in 2014. LTE has more capacity available so MBMS may be more successful there.

Release 7 was the last 3G-only release before the arrival of LTE. The most important enhancements in this release include:

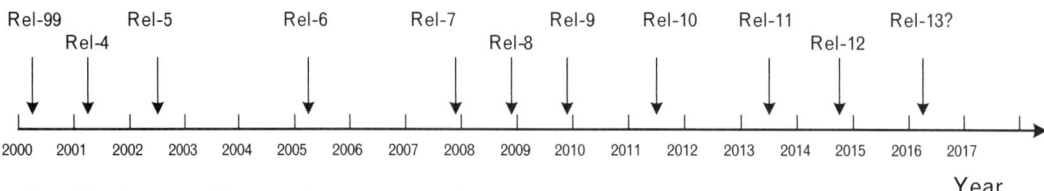

FIGURE 2.6 Timeline of 3GPP releases.

- HSPA+;
  - Advanced mobile receiver diversity;
  - Higher order modulation (64QAM for the downlink, and 16 QAM for the uplink);
  - Continuous packet connectivity (CPC);
  - Multiple input multiple output (MIMO) antennas;
  - RAN architecture improvements (flat architecture, lower latency);
- Evolved EDGE.

The maximum data rate in Release 7 is 21.2 Mbps for the downlink, and 11.5 Mbps for the uplink.

Releases 8 and later contain also LTE; therefore those releases are discussed in Chapter 10. See Table 2.6 for a timeline of 3GPP releases.

2.3.3 3GPP2

2.3.3.1 Introduction

The 3GPP2 initiative is the other major 3G standardization organization. It promotes the code division multiple access 2000 (CDMA2000) system, which is also based on a form of WCDMA technology. In the world of IMT-2000, this proposal is known as IMT-MC. The major difference between the 3GPP and the 3GPP2 approaches is that 3GPP has specified a completely new air interface without constraints from the past, whereas 3GPP2 has specified a system that is backwards compatible with IS-95 systems. This approach was seen necessary because in North America, IS-95 systems already use the frequency bands allocated for 3G by the World Administrative Radio Conference (WARC). It makes the transition into 3G much easier if the new system can coexist with the old system in the same frequency band. The CDMA2000 also uses the same core network as IS-95, namely IS-41 (also known as TIA/EIA-41).

A CDMA2000 system can operate on the same 1.25-MHz radio channels as IS-95, and thus it offers backward compatibility with IS-95. The chip rate of CDMA2000 (1x) is also the same as in IS-95 (1.2288 Mcps). Note that the CDMA2000 standard originally included two carrier modes; single-carrier (1x) and multi-carrier (3x) modes (see Figure 2.7). However, the 3x mode never took off as intended (that is, as a wideband carrier system with 3.75 MHz carriers), and all CDMA2000 deployments so far are using the single carrier mode only.

2.3.3.2 1xRTT

The first CDMA2000 standard was called 1xRTT. This abbreviation stands for single carrier radio transmission technology. Whether or not it is a true 3G technology depends on who does the definition. The theoretical maximum data rate in the first release (cdma2000 revision 0) was 153 Kbps, though in operational networks this is likely to remain below 100 Kbps. After revision 0, several new revisions have been released: revision A, revision B, revision C, revision D, and revision E, which is also known as 1x advanced (more about 1x Advanced in Section 2.3.3.5). The first cdma2000 1xRTT network was deployed in October 2000.

2.3.3.3 1xEV-DO

CDMA2000 1xEV-DO (evolution-data optimized) is a family of standards that brought new high-speed packet-switched techniques to CDMA2000. The design aim was to deliver peak data rates exceeding 2 Mbps. EV-DO introduces an additional carrier that is optimized for packet data transfer only. In an EV-DO system the circuit-switched connections (voice) and packet-switched high-speed data connections are separated on different 1.25 MHz carriers, even if they belong to the same user. Indeed, originally EV-DO abbreviation meant "evolution-data only".

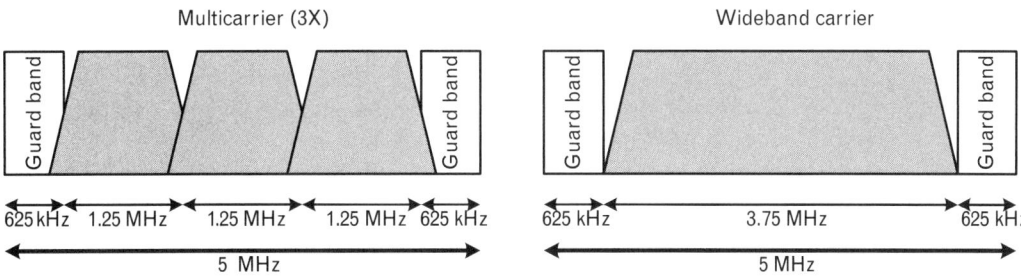

FIGURE 2.7 *Single-carrier versus multicarrier modes.*

1xEV-DO has also undergone several revisions since its first release (release 0): revision A and revision B. There were also plans for revision C, which would have included an OFDMA-based 4G air interface. However, as discussed later, this revision will not be deployed.

EV-DO release 0 offers data rates up to 2.4 Mbps in the downlink (or in the forward link in CDMA2000 jargon), which in operational networks means approximately 300–600 Kbps. This release has many similarities with HSDPA, which is part of UTRAN Release 5. The maximum uplink data rate is 153 Kbps. The first EV-DO release 0 network was deployed in 2002.

EV-DO revision A introduces further enhancements by reducing latency and increasing data rates. The maximum downlink data rate is increased to 3.1 Mbps, and the uplink rate to 1.8 Mbps. Low latency (around 50 ms) enables new types of services such as VoIP and video telephony. EV-DO rev A has many similarities with HSUPA, which is part of UTRAN Release 6. The first EV-DO rev A network was deployed in 2006.

EV-DO revision B introduces scalable bandwidth channels. Revision B enables the aggregation of multiple 1.25 MHz channels to form a set of parallel channels, increasing the data throughput. The standard allows 15 parallel channels, in which case the combined channel has an effective bandwidth of 20 MHz. Note that the component channels do not have to be adjacent in frequency space. Also, with the introduction of 64QAM modulation, the maximum data rate of a single 1.25-MHz component carrier has been increased to 4.9 Mbps. An aggregated 5-MHz carrier can thus deliver 14.7 Mbps downlink, and a 20 MHz carrier could achieve 73.5 Mbps downlink. However, these numbers are truly theoretical because 64QAM requires very good radio conditions. Latency is also further reduced in EV-DO revision B. The first EV-DO revision B network was deployed in early 2010.

2.3.3.4 EV-DO Advanced

EV-DO advanced is the latest enhancement to EV-DO systems. It does not improve the actual theoretical maximum data rate. Rather, it is a software update that improves network operation and management processes. EV-DO advanced enhancements include the following:

- Network load balancing (NLB) reassigns mobile devices from heavily loaded sectors to lightly loaded sectors, even if the radio conditions are better in the heavily loaded sector.

- Adaptive frequency reuse reduces the interference in a capacity-constrained cell by reducing the transmit power of a secondary carrier in an adjacent lightly loaded cell, while the primary carrier continues to operate at full power to ensure wide coverage. As a result, the overall network capacity is increased.

- Distributed network scheduler prioritizes and allocates bandwidth to multiple users at the network level to help to maximize overall network efficiency.
- Single carrier multilink, whereby two carriers using the same frequency, can serve a dual-antenna mobile device from different cells or sectors. Therefore the mobile can get the benefits of a multicarrier network, even though it moves in a single carrier network.
- Smart carrier management can assign one or more carriers to a device within the same cell or from an adjacent cell if another cell is less loaded or offers better coverage.
- Enhanced connection management optimizes device connections to the network depending on the type of application being used to improve signaling capacity and to reduce latency.

2.3.3.5 1x Advanced

Whereas EVDO enhancements were about improving the packet switched operations of the CDMA2000 system, 1x advanced is a standards release for improving the circuit-switched side of the system. The specific 1x advanced enhancements include the following:

- Enhanced variable rate codec (EVRC-B);
- Quasi-linear interference cancellation (QLIC);
- Quasi-orthogonal functions (QOF);
- Advanced QLIC;
- Mobile receive diversity (MRD);
- Radio link enhancements, including:
  - Reverse link interference cancellation (RLIC);
  - Smart blanking;
  - Efficient power control;
  - Frame early termination (FET).

As a result, 1x advanced can increase the voice capacity in the system by up to a factor of four, or alternatively to increase network coverage by up to 70 percent. The peak data rate is increased to 307 Kbps.

26 HISTORY OF MOBILE TELECOMMUNICATIONS

2.3.3.6 Evolution Data and Voice (1xEVDV)

This evolution was to enhance the existing systems for data and voice on the same carrier. Note that as the name says, EVDO systems are optimized for data, and 1x advanced is designed for voice. Therefore, in state-of-the-art CDMA2000 systems, data and voice tend to use different carriers. However, due to lack of interest from network operators, Qualcomm halted the development work of 1xEVDV in 2005.

2.3.3.7 Ultra Mobile Broadband (UMB)

This system was planned to be 3GPP2's version of 4G. It was to be based on OFDMA, in a similar way as LTE. However, in 2008 Qualcomm decided to support LTE as the 4G technology instead. This means that while CDMA2000 keeps on evolving, there will not be major upgrades to the system anymore. The upgrade path for CDMA2000 operators will point to LTE.

2.4 Conclusions

In this chapter we have quickly reviewed the history of mobile telecommunications. Timewise, this history is surprisingly short. The first generation systems were launched about 30 years ago. During those 30 years the mobile telecommunications world has undergone not an evolution but a revolution. Nobody in early 1980s could have forecast the way mobile telecommunications services are used today. The general public has embraced mobile phones with such enthusiasm that it has surprised the system designers too. All generations so far (1G, 2G, and 3G) have suffered from the shortage of capacity since they were designed for fewer users or less data. See Figure 2.8 for a graph of mobile communications generations and the technologies in them.

The fourth generation, or LTE-Advanced, tries to meet the challenges of today and those of the future. It is a continuously evolving standard, with each release containing new, improved features. It is good to point out that basic LTE (without the "A" part) as such is not a recognized 4G system, and some call the basic LTE a 3.9G system. LTE was first introduced in 3GPP Release 8, but it did not meet the requirements for a 4G system set by ITU. It is only in Release 10, which was called LTE-Advanced, that those requirement could be met. At the time of this writing, Release 11 specifications are available and frozen, and work is undergoing with Release 12.

It is also interesting to note that so far new mobile communication generations have been launched about every 10 years. The first generation started around 1980, the second after 1990, the third generation had its first

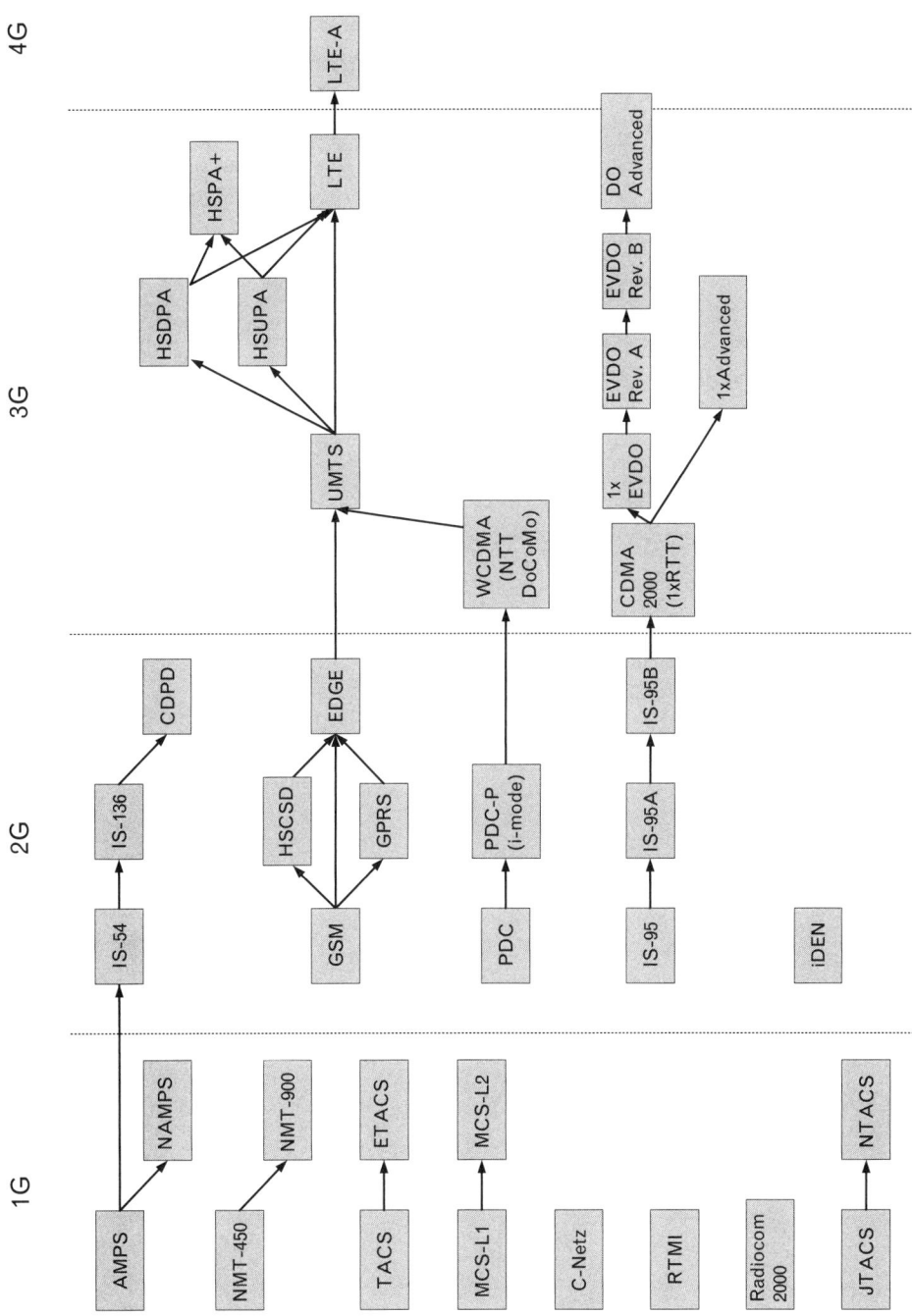

FIGURE 2.8 *Mobile communications generations and technologies in them.*

deployments around the millennium, and the fourth generation networks have been around since 2010. There is no reason why this trend could not continue, with a fifth generation system being launched in 2020. However, it is good to note that the definition of a new generation is not very clear cut. Great improvements in capacity, performance, and services can also be achieved within a generation, using the existing basic technology.

References

[1] Watanabe, K., and K. Iwamura, "Evolution of NTT High-Capacity Land Mobile Communication System," *IEEE International Conference on Communications,* Boston, MA, 1989.

[2] Redl, S., M. Weber, and M. Oliphant, *An Introduction to GSM,* Norwood, MA: Artech House, 1995.

[3] Korhonen, J., *Introduction to 3G Mobile Communications,* Norwood, MA: Artech House, 2003.

CHAPTER 3

Overview of a Modern 4G Telecommunications System

3.1 Introduction

The LTE-advanced is now de facto 4G mobile communications system, and it is likely to remain so. It does not have a serious competitor in sight. 3GPP2 has given up its UMB initiative, and mobile WiMAX (802.16m) has not been able to gain significant market share, though it fulfills the 4G criteria set by the ITU.

In order to understand LTE-A, it is better to have a look at LTE first. LTE is a very different system from UMTS, as both its architecture and the technologies used are either new or greatly enhanced versions of the old entities. In this section we will first have a look at LTE architecture, and then we will discuss the radio access network (RAN), the new OFDM air interface, and the core network. The final section presents the LTE-A enhancements to LTE. The purpose of this chapter is to give an overview of an LTE-A system. The detailed discussion will then follow in subsequent chapters.

3.2 LTE-A System Architecture

Figure 3.1 gives a high-level description of the LTE-A network architecture. Readers who are more familiar with 2G/3G networks may notice the simplicity of the LTE-A architecture. In the old GSM there were base transceiver stations (BTS) and base station controllers (BSC), and in UTRA networks (see Figure 3.2) we have NodeBs and radio network controllers (RNC), and several different entities in the core network.

30 OVERVIEW OF A MODERN 4G TELECOMMUNICATIONS SYSTEM

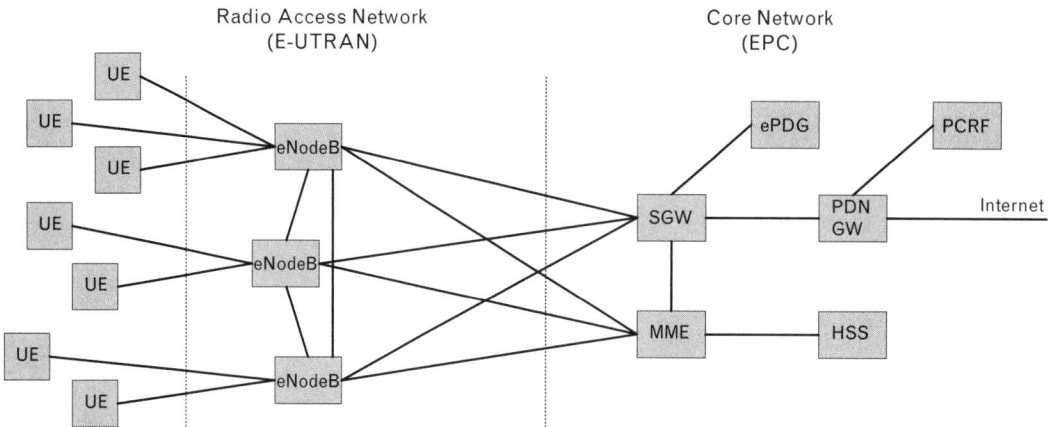

Figure 3.1 *LTE system architecture.*

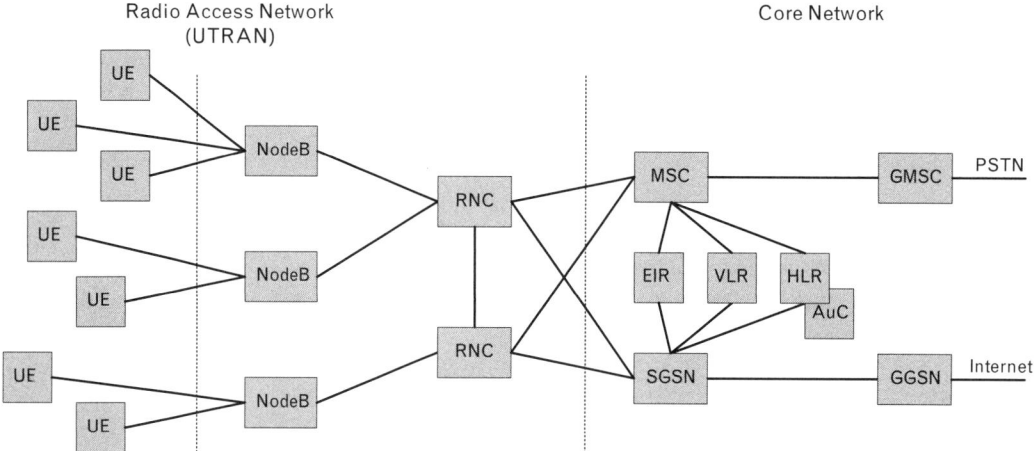

Figure 3.2 *UTRAN system architecture.*

3.3 LTE RAN

In LTE we have only eNodeBs in the RAN network, though eNodeBs come in different sizes and configurations. There is no separate controller element like in GSM or in UTRAN. This change is mainly because of the drive for lower response times. Latency is an important quality of service (QoS) parameter. Many interactive applications such as interactive games require very short response times, and the more network elements there are on the signal path, the more time will be spent on processing the signal before it reaches its destination. Also, by bringing more intelligence to the eNodeB, it may be able to handle more requests from the mobile directly, without consulting the core network. This will result in shortened latencies for control signaling.

The third generation mobile communication system as defined by 3GPP had a WCDMA-based air interface. Whereas WCDMA has many good qualities, the requirements for a state-or-the-art mobile communication system have changed, and an OFDM-based system can fulfill these better. Therefore, OFDM was adopted as the air interface technology for LTE RAN. In the following section we will discuss why that is so.

3.4 OFDM Air Interface

First of all, OFDM is a system where the wideband carrier consists of numerous subcarriers. In WCDMA a single wideband carrier is employed, and it can be very wide indeed. This is probably the most important difference between these systems, and it provides several advantages to an OFDM system over a WCDMA system. The channel response (i.e., how the channel distorts the signal that passes through the channel) depends on the frequency used. Therefore, with wideband carriers the channel response is different in different parts of the frequency channel. Such a channel is called a frequency-selective channel, and designing a receiver for a frequency-selective channel is challenging. When OFDM is employed, a wideband carrier is transformed into many narrowband carriers, which are frequency-flat (i.e., the channel response is the same for the whole carrier), and receivers for frequency-flat channels are easier to design. The list of OFDM advantages over WCDMA include:

- Operators have found out that it is difficult to acquire spectrum, and it is especially difficult to get spectrum that is in one big, continuous block. A high-capacity mobile communications system needs more spectrum since in principle more spectrum equals more capacity. However, a WCDMA carrier has to be continuous in frequency, and that is a serious limitation for many operators. In OFDM a carrier can consist of component carriers, and those do not have to be continuous in frequency. Therefore, an operator can use even small frequency bands and combine them to larger carriers in OFDM. Carrier bandwidths in OFDM vary between 1.4 MHz and 20 MHz.

- LTE-A has introduced a concept of carrier aggregation, which enables the network to combine up to five component carriers to form one wideband composite carrier. Since the maximum bandwidth of a component carrier is 20 MHz, carrier aggregation could handle up to 100 MHz composite carrier bandwidths. Note that it is true that late

WCDMA specifications also allow multicarrier deployments. In Release 10 an HSDPA system can assign four normal 5-MHz carriers to a single user. However, multicarrier HSDPA still cannot complete with the level of flexibility provided by an OFDM system.

- OFDM is better at dealing with multipath interference. Since WCDMA employs (possibly very) wide carrier bandwidths, it is difficult for the receiver to equalize over the whole carrier. OFDM is a multicarrier system; the total carrier bandwidth consists of several narrowband subcarriers, and thus equalization can be done individually for each such carrier.

- OFDM is better at providing a single frequency network. Such networks are useful for multimedia broadcast multicast (MBMS) services. Single frequency networks may suffer from intersymbol interference on cell boundary areas. However, since OFDM divides the transmission bandwidth into a large number of narrowband subcarriers, the resulting symbol length is quite long, and the guard interval between the symbols ensures that the symbols do not interfere with each other at the receiver.

- WCDMA cells are interference-limited (i.e., their capacities depend on interference caused by other users). From this fact follows that the effective cell size is also changing according to a number of active users in the cell. This can result in a phenomenon called cell breathing, whereby a highly loaded cell loses coverage, thus also interfering users. And due to less interference, the effective cell size grows again. This phenomenon can be alleviated by careful transmission power control and load balancing, but it will still be a problem that needs attention. OFDM cells are not interference-limited in a similar way as WCDMA cells, and thus cell breathing does not exist.

- OFDM is particularly suitable for MIMO. OFDM modulation turns a frequency-selective MIMO channel into a set of parallel frequency-flat MIMO channels. As a result, the MIMO multichannel equalization will be simpler, since only a constant matrix has to be inverted for each OFDM subcarrier.

- Higher spectrum efficiency is possible with MIMO than what WCDMA-MIMO can achieve. For 2x2 downlink MIMO, the numbers are about 0.6 bps/Hz in WCDMA, 1.7 bps/Hz in LTE, and 2.4 bps/Hz in LTE-A [1].

OFDM as a technology is further discussed in Chapter 4. The previous bullet points list only the differences that make it more suitable than WCDMA for a 4G system. However, it is good to point out that WCDMA is by no means a dying technology in mobile communications. UTRA networks are being enhanced in every 3GPP release, and, for example, a Release 12

UTRAN system is certainly a much more capable system than a Release 6 system, even though both are based on WCDMA.

3.5 Evolved Packet Core

3GPP introduced LTE in Release 8, and it is best known for its new air interface technology, OFDM. However, the air interface was not the only new system component that was introduced in Release 8, since also the core network was new. The new core network architecture is known as the system architecture evolution (SAE), or more commonly as the evolved packet core (EPC). EPC is based on the GPRS core network, and it uses the Internet protocol (IP) as the basic transmission protocol, which means that there is no circuit-switched domain left in the system, and only the packet-switched domain exists.

The main design principle in EPC was to keep the architecture simple. In GPRS and UMTS the network has a hierarchical architecture with many different types of nodes, which also results in a large number of different interfaces between them. A signal/message is likely to be processed in many nodes and sent over many interfaces, resulting in unnecessary long latencies for message exchanges. To simplify things, EPC provides a flat architecture, with a few new network nodes.

The user plane (i.e., the data traffic) and the control plane (control signaling) are logically separated. The architecture is optimized for the data traffic—the aim is to transfer the data as quickly as possible, with as few network nodes and protocol conversions on the data path as possible.

The network nodes in the EPC include serving gateway (S-GW), packet data network gateway (PDN GW), mobility management entity (MME), and home subscriber server (HSS). If the radio access network used is not a 3GPP RAN, then some other types of nodes may also be needed in the core network.

As with the air interface, EPC is constantly evolving, and new improvements are introduced in every release. EPC will be discussed further in detail in Chapter 8.

3.6 LTE Requirements

Quite soon after the first 3GPP networks were launched, it became obvious that 3G had a problem with data transfer services. Though 3G can transfer data, it cannot transfer a lot of it. With the increasing number of data applications, the capacity of 3G would run out sooner rather than later. There are ways to increase capacity (e.g., by adding more spectrum), but WCDMA is not very good at very wide carrier bandwidths.

3GPP held a workshop in late 2004 in Toronto, and, based on the contributions presented, a feasibility study regarding a new system was launched. The outcome of this work was Technical Report 25.913 [1], which contained a list of requirements for the future LTE system. When looking at this list, it becomes clear that many of these requirements would be difficult to achieve merely by enhancing the existing UMTS system. In the following sections we will present and discuss these requirements.

LTE requirements for Release 8 [1] are listed as follows:

- Peak data rate:
  - Instantaneous downlink peak data rate of 100 Mbps within a 20-MHz downlink spectrum allocation (5 bps/Hz).
  - Instantaneous uplink peak data rate of 50 Mbps within a 20-MHz uplink spectrum allocation (2.5 bps/Hz).
- Control-plane latency:
  - Transition time of less than 100 ms from a camped state, such as Release 6 idle mode, to an active state such as Release 6 CELL_DCH.
  - Transition time of less than 50 ms between a dormant state such as Release 6 CELL_PCH and an active state such as Release 6 CELL_DCH.
- Control-plane capacity:
  - At least 200 users per cell should be supported in the active state for spectrum allocations up to 5 MHz.
- User-plane latency:
  - Less than 5 ms in unloaded condition (i.e., single user with single data stream) for small IP packet.
- User throughput:
  - Downlink: average user throughput per MHz, 3 to 4 times of Release 6 HSDPA.
  - Uplink: average user throughput per MHz, 2 to 3 times of Release 6 enhanced uplink.
- Spectrum efficiency:
  - Downlink: In a loaded network, target for spectrum efficiency (bits/sec/Hz/site), 3 to 4 times of Release 6 HSDPA.

- Uplink: In a loaded network, target for spectrum efficiency (bits/sec/Hz/site), 2 to 3 times of Release 6 enhanced uplink.

- Mobility:
  - E-UTRAN should be optimized for low mobile speed from 0 to 15 km/h.
  - Higher mobile speed between 15 and 120 km/h should be supported with high performance.
  - Mobility across the cellular network shall be maintained at speeds from 120 km/h to 350 km/h (or even up to 500 km/h depending on the frequency band).

- Coverage:
  - Throughput, spectrum efficiency, and mobility targets listed earlier should be met for 5-km cells and with a slight degradation for 30-km cells. Cells range up to 100 km should not be precluded.

- Further enhanced multimedia broadcast multicast service (MBMS):
  - While reducing terminal complexity: same modulation, coding, multiple access approaches and UE bandwidth than for unicast operation.
  - Provision of simultaneous dedicated voice and MBMS services to the user.
  - Available for paired and unpaired spectrum arrangements.

- Spectrum flexibility:
  - E-UTRA shall operate in spectrum allocations of different sizes, including 1.25 MHz, 1.6 MHz, 2.5 MHz, 5 MHz, 10 MHz, 15 MHz, and 20 MHz in both the uplink and downlink. Operation in paired and unpaired spectrum shall be supported.
  - The system shall be able to support content delivery over an aggregation of resources including radio band resources (as well as power, adaptive scheduling, and so on) in the same and different bands, in both uplink and downlink and in both adjacent and nonadjacent channel arrangements. A "radio band resource" is defined as all spectrum available to an operator.

- Co-existence and interworking with 3GPP radio access technology (RAT):
  - Co-existence in the same geographical area and co-location with GERAN/UTRAN on adjacent channels.

- E-UTRAN terminals supporting also UTRAN and/or GERAN operation should be able to support measurement of, and handover from and to, both 3GPP UTRAN and 3GPP GERAN.
- The interruption time during a handover of real-time services between E-UTRAN and UTRAN (or GERAN) should be less than 300 msec.

- Architecture and migration:
  - Single E-UTRAN architecture.
  - The E-UTRAN architecture shall be packet based, although provision should be made to support systems supporting real-time and conversational class traffic.
  - E-UTRAN architecture shall minimize the presence of "single points of failure."
  - E-UTRAN architecture shall support an end-to-end QoS.
  - Backhaul communication protocols should be optimized.

- Radio resource management requirements:
  - Enhanced support for end-to-end QoS.
  - Efficient support for transmission of higher layers.
  - Support of load sharing and policy management across different radio access technologies.

- Complexity:
  - Minimize the number of options.
  - No redundant mandatory features.

Peak data rate is a popular attribute among marketing people and the press. This rate is typically so high that it sounds good for prospective customers. However, it is really a theoretical peak data rate in ideal conditions, and as such it will not be available in operational networks. It is also the maximum capacity of the whole base station, or a base station sector, and therefore it has to be divided between the users of the base station/sector. And even if the user were the only user of the base station in question, a standard mobile device could not make use of the whole available capacity.

Peak data rate can be increased by adding more spectrum, by using higher-order modulation, or by MIMO antenna techniques. Here, the peak data rate requirement is already given as "per Hertz" so simply adding more spectrum to the system would not help. But LTE employs both MIMO and higher-order modulation. It supports 4x4 MIMO antennas in the downlink and 2x2 MIMO in the uplink in Release 8. The modulation schemes

FIGURE 3.3
Latency improvement in LTE.

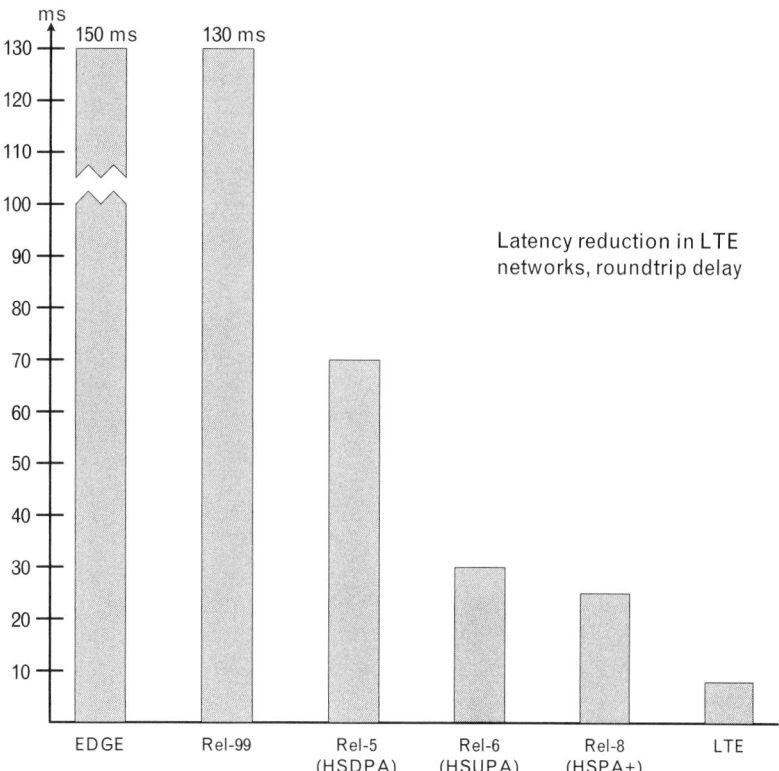

supported are QPSK, 16QAM, and 64QAM (64QAM is optional for the UE in the uplink). With these parameters, the peak data rate will easily exceed the requirements both in the uplink and in the downlink [2].

Control-plane latency requirements between idle/dormant and active states for the mobile device are 100 ms between idle and active states, and 50 ms between dormant and active states. This requirement has been fulfilled mainly by adopting a simplified and flat architecture, since this will reduce the transmission delays. The new eNodeB also includes most the functionality of the old base station controller node (RNC in UTRAN). Since the eNodeB has more intelligence, it can itself respond to more queries from the mobile station, without consulting other network nodes first. However, control plane latency is a complex issue, and low latency requires careful planning of control protocols. Simply reducing the number of messages transferred will also reduce the latency caused by the whole procedure. The signaling procedures should be as simple as possible, and the transfer of redundant information (i.e., information that is already known by the mobile station or information which is not needed) should be avoided. Sometimes the UE also needs to read the broadcast system information, and the system information transmission sequences should be designed so that the most commonly needed information is transmitted more often.

Control-plane capacity requirement is that at least 200 users per cell should be supported in the active state for spectrum allocations up to 5 MHz. This requirement (as given in [1]) is a bit unclear, but if interpreted that the system should be able to support this number of users per cell with quasi-instantaneous access (i.e., within 1 ms) to radio resources in the active state, then it can be achieved: one resource block is 180 kHz wide and 0.5 ms in duration. A 5-MHz-wide channel can accommodate 27 such blocks in the frequency domain. This number can be multiplied by two because an LTE transmission time interval (TTI) (1 ms) can take two resource blocks (of 0.5 ms length) => 54 users in each TTI. Furthermore, if 4x4 MIMO is used, then this number can be further multiplied by 4 => 216 simultaneous users.

User-plane latency should be less than 5 ms in unloaded conditions for a small IP packet (i.e., with no payload). Small IP packet means that the message does not need to be segmented and can be sent in a single subframe. In the air interface the most important factor deciding the amount of latency is the TTI. TTI defines the duration of a transmission on the air interface. In LTE the TTI is only 1 ms. That is, it will take 1ms for the UE to transmit even a short data packet. In addition there will be small processing delays in the mobile station, eNodeB, and in the S-GW, plus a transmission delay between the eNodeB and the S-GW, which depends on the length of the cable between these nodes (i.e., their physical distance). In any case 5 ms is achievable in an unloaded network where retransmissions are not needed.

User throughput/spectrum efficiency requirement was given as improvement over Release 6; in the downlink LTE should be 3–4 times better than Release 6 HSDPA, and in the uplink it should be 2-3 times better than Release 6 enhanced uplink. These are also achievable, provided that 2x2 MIMO is used in the downlink and 1x2 MIMO in the uplink.

Mobility requirements are divided into three classes: the emphasis in LTE performance should be for low mobile speeds from 0 to 15 kph. Still, mobile speeds between 15 and 120 kph should be supported with high performance. And some level of mobility should be maintained up to 350 kph (or even up to 500 kph, depending on the frequency band). Note that these kinds of requirements are difficult to measure. For example, who defines what is the "high performance" that should be supported for mobile with velocity between 15 and 120 kph?

The coverage requirement states that throughput, spectrum efficiency, and mobility targets should be met for 5-km cells, and with a slight degradation for 30-km cells. Cells range up to 100 km should not be precluded. Again, the requirements for 30-km and 100-km range cells are not exact and therefore cannot be measured.

The requirement for further enhanced multimedia broadcast multicast service (MBMS) is included in the requirements list, but in the end this feature was not included in Release 8 LTE. The reason for this was probably the extra work required and the fact that 3G-based MBMS was not deployed commercially anywhere in the world and thus there was no urgent need to specify it for LTE from the start. However, eMBMS is included in LTE Release 9.

The spectrum flexibility requirement is actually not a requirement but a relief. LTE is very flexible when it comes to spectrum bands. Whereas UTRAN could only operate on 5-MHz carriers (in the FDD mode that is; the TDD mode has more options), LTE can use six different carrier bandwidths: 1.4, 3, 5, 10, 15, and 20 MHz. Both paired and unpaired spectrum can be used. Reference [1] also mentions carrier aggregation as a means to combine these component carriers into bigger aggregated carriers. The component carriers do not have to be on adjacent frequency bands. However, carrier aggregation is not part of Release 8, but it is introduced in Release 10.

Co-existence and interworking with 3GPP radio access technology (RAT) means that LTE and GERAN/UTRAN networks can have overlapping coverage areas, and an LTE mobile device should be able to support measurements of, and handover from and to, both UTRAN and GERAN. For real-time services, the interruption time during handover should be less than 300 ms; for non–real time services this requirement is less than 500 ms.

The requirements for architecture and migration do not contain many tangible details, except that there should be only one E-UTRAN architecture (which is common sense) and that architecture should be packet based.

Similarly, the requirements for radio resource management are not really measurable; rather, they form a list of good intentions. Load sharing across different RATs is an especially useful feature.

The complexity requirement is also recommendable. The number of options and features should be kept at minimum. In early 3GPP work, if competing features were proposed for addition to specifications, it was sometimes easiest to add them all as optional features. This approach will result in unnecessary complex and expensive devices, and it is especially difficult to test all those optional features.

This list of requirements here is not complete; the reader can have a look at [1] for the full list. However, the most important requirements are listed here; the rest are mainly good intentions of what the system should be able to do.

As a conclusion, it is clear that a new system was needed, and the new system should be able to provide much higher data rates to many more users. They key improvements to achieve these were the new system architecture and the selection of OFDM as the new air interface technology. OFDM can make use of very wide frequency bands efficiently, especially when carrier aggregation is used. It is also well suited for MIMO, which is definitely a technology to be exploited when higher data rates are needed. Most of all, it was important to have a system that can be enhanced in the future. Future-proof is not quite the right word because no system can foresee and accommodate every possible future requirement. However, the nature of 3GPP work is such that new features will be introduced in every release, and LTE provides a good platform to do it. LTE-advanced is already a proof of this.

3.7 LTE-Advanced

LTE was in first introduced in 3GPP Release 8, which was therefore a major release. After that, Release 9 was a more lightweight release for LTE, containing features that were originally planned for Release 8 and for some reason or another were postponed.

3GPP Release 10 was again a major release, and it included the standards for LTE-advanced. Even though the basic LTE was a great improvement over UMTS, it still fell short of the requirements set by ITU for a 4G system. LTE-advanced aimed to fix this.

This section explains the issues that make LTE-advanced different from the standard Release 8 LTE. In fact, the transition from LTE to LTE-A is much smaller than the transition from UMTS to LTE. LTE already included the major changes, such as the new air interface based on OFDMA.

ITU has defined a set of high-level requirements that have to be fulfilled before a system can claim to be a 4G system (or as ITU defines it, IMT-Advanced). These are given in the following list [2]:

- A high degree of common functionality worldwide while retaining the flexibility to support a wide range of services and applications in a cost-efficient manner;
- Compatibility of services within IMT and with fixed networks;
- Capability of interworking with other radio access systems;
- High-quality mobile services;
- User equipment suitable for worldwide use;
- User-friendly applications, services, and equipment;
- Worldwide roaming capability;
- Enhanced peak data rates to support advanced services and applications (100 Mbps for high and 1 Gbps for low mobility were established as targets for research).

As one can see, the problem with these requirements is that, except for the last requirement, they are rather difficult to measure. Any system can claim to offer high-quality mobile services and user-friendly applications. The last requirement looks rather challenging at first, but it is good to note that this is about peak data rates. Again, peak data rates are often purely theoretical values, not to be achieved or maintained in operational networks. Therefore simply by throwing in a few enhanced parameters and making assumptions, the peak data rate can be increased considerably. For example, with MIMO 8x8 antennas, higher-order modulation, and very wide channel bandwidths, it is possible to show that the new system can achieve 1Gbps data rate in ideal conditions. But will you get this data rate to your mobile device in five years' time? The answer is certainly no.

In fact, it is more interesting to analyze 3GPP's own characterization of the LTE-A system from Annex 1 of [1, 3]. This list contains lots of hard facts that make it interesting reading for a telecommunications engineer:

- Downlink transmission scheme is based on conventional OFDM;
- Uplink transmission scheme is based on discrete Fourier transform (DFT) spread OFDM;
- Both FDD and TDD operations are supported;
- Transmission bandwidths up to 100 MHz are supported, individual carriers up to 20 MHz are supported, and a maximum of five of them can be combined with carrier aggregation;
- Peak data rates of 1Gbps in the downlink and 500 Mbps in the uplink (this was later increased to 3 Gbps in the downlink and 1.5 Gbps in the uplink);
- Channel coding is based on rate 1/3 turbo coding + hybrid ARQ with soft combining;
- Data modulation supports QPSK, 16QAM, and 64QAM, for both the downlink and the uplink;
- 8x8 MIMO supported on the downlink, 4x4 on the uplink;
- Intercell interference coordination (ICIC) is supported;
- Relays are supported;
- Flat radio access network (RAN) architecture (RAN has only one type of node, eNodeB);
- Control plane latency 50 ms;
- User plane latency 4 ms;
- Peak spectrum efficiency; downlink 30 bps/Hz, uplink 15 bps/Hz;
- Mobility support for speeds up to 350 kph (possibly up to 500 kph).

Many of these requirements were already supported by Release 8/9 LTE. The completely new LTE-A features include:

- Carrier aggregation;
- Higher order MIMO;
- Relays;
- Enhanced intercell interference coordination (eICIC);
- MBMS enhancements;
- SON enhancements.

With the help of carrier aggregation and higher order MIMO, it is possible to achieve the new peak data rate requirements. And rather than calculating peak spectrum efficiency, it is more useful to examine the average spectrum efficiency requirement. Peak spectrum efficiency means error-free conditions and all radio resources of a cell assigned to a single mobile device. This kind of situation will never happen in operational networks. Average spectrum efficiency is defined as the aggregate throughput of all users (the number of correctly received bits over a certain period of time) normalized by the overall cell bandwidth divided by the number of cells. This gives a much better estimate of the performance of the network, though it is good to remember that these numbers are also based on computer simulations and not on measurements in deployed networks.

For LTE-A the targets for average spectrum efficiency are listed in Table 3.1.

In Table 2.1, the antenna configuration column gives the antenna configuration employed. For example, DL 4x2 refers to a downlink antenna system that has four transmitting and two receiving antennas. And since this is downlink, the four transmitting antennas are in the base station, and the two receiving antennas are in the UE.

Carrier aggregation was introduced into LTE-A in Release 10. It is used to combine several component carriers (up to five) into one logical

TABLE 3.1 LTE-A TARGETS FOR AVERAGE SPECTRUM EFFICIENCY

ANTENNA CONF.		AVERAGE THROUGHPUT [BPS/HZ/CELL]
UL	1x2	1.2
	2x4	2.0
DL	2x2	2.4
	4x2	2.6
	4x4	3.7

aggregated carrier, though in Releases 10 and 11 only two-component carriers are supported. The component carriers are normal carriers (i.e., similar to standard LTE carriers), and thus they can have a bandwidth of 1.4, 3, 5, 10, 15, or 20 MHz. Therefore, the theoretical maximum bandwidth of an aggregated carrier is 100 MHz. However, it is very unlikely that such a wide carrier is seen in operational networks (first of all, such large unused frequency blocks simply do not exist, and, even if one was made available, no operator would be wealthy enough to buy it). Rather, the main reason for carrier aggregation is that some operators may have access to only relatively small slices of spectrum on different frequency bands, and carrier aggregation provides a way to combine those slices into wider, more useful, spectrum channels. Carrier aggregation is discussed in detail in Chapter 11.

MIMO was also enhanced in Release 10. The downlink now supports 8x8 MIMO, and in the uplink 4x4 MIMO can be used. This kind of higher-order MIMO system can increase the data throughput considerably, but it must be noted that 8x8 MIMO requires very good radio conditions before it can be used successfully. However, MIMO as a technology is much more than just adding more antennas to devices, and it is discussed in detail in Chapter 11.

Relays are the third important LTE-A enhancement. The reason to introduce relays is that macro-cells may have locations with poor coverage, especially near cell edges. Relays offer an inexpensive way to cover those locations better. A new macro eNodeB would cost more, and the wired backhaul especially may become a costly issue. LTE-A relays employ wireless backhaul, which is in principle free. However, it uses the same spectrum as other base stations/users, and thus the capacity of base stations involved is less than it would be without relays. LTE-A relays are so-called layer 3 relays, which are transparent to mobile devices (i.e., mobile devices do not know that the "base station" they are connected to is not a base station but a relay since the relay looks like a normal base station to mobile devices). The backhaul may use the same or a different frequency compared to the frequency used by the relay. Relays are also discussed in detail in Chapter 11.

Enhanced intercell interference coordination (eICIC) is introduced to deal with interference, especially in heterogenous networks (HetNet). Pico and femto cells are becoming more popular, and they are introduced especially to traffic hotspots. However, these cells are low power, and they may suffer from serious interference from the much higher power macro cells, since typically such pico cells and macro cells have overlapping coverage. The solution proposed in Release 10 is to use almost blank subframes (ABS). During those frames, only the most minimal set of signals is transmitted from the macro eNodeB (or whoever is the interferer). The pico/femto cells can then use these quiet frames to transmit to the UEs in their coverage area. More about eICIC can be found in Chapter 11.

MBMS is also enhanced in Release 10. MBMS was already introduced in LTE in Release 9, and the most important enhancement in Release 10 is the counting function. Because MBMS is a broadcast/multicast service, the network does not necessarily know how many users (if any) are receiving a particular broadcast. This may lead to a waste of resources if the network broadcasts services nobody wants to receive. The counting function checks the number of connected mode users that either already receive, or want to receive, a particular MBMS service. The details of this feature and MBMS in general are discussed in Chapter 11.

Self-organizing networks (SON) enhancements is a very diverse topic. SON is an "umbrella" work item for various issues that aim to assist network operators in managing LTE networks. SON involves various mechanisms that maintain or improve the performance of LTE network autonomously without an operator's intervention, or at least with reduced need for interventions. SON is such a wide topic that it is discussed in its own section in Chapter 11.

The first 3GPP release implementing LTE-A feature was Release 10. Release 10 was frozen (i.e., no new functional features can be added to the specification after this date) in March 2011, and in 2013 there are already some network operators trialing these features. It is also good to point out that the new LTE-A features presented in this section are only a subset of all new LTE-A features (albeit they are they are the most important ones).

3.8 LTE-A in Release 11

Release 11 was nominally frozen in September 2012, but in practice the RAN protocols were stable only in March 2013.

The list of LTE-A enhancements and new features include:

- Coordinated multipoint transmission and reception (CoMP);
- Carrier aggregation enhancements;
- Further enhanced intercell interference coordination (FeICIC);
- MBMS service continuity;
- Further SON enhancements.

CoMP is a scheme where the data throughput for UEs near cell edges is improved, and thus the overall system efficiency is enhanced. The idea behind CoMP is that the UEs near cell edges can receive the signal from two or more cells, and conversely such UEs' transmissions can be received by several cells. If the downlink transmissions are coordinated, and the same data is sent to the UE via several transmitters, this obviously improves the quality of the reception. There are several variations in this scheme, and they are all discussed further in Chapter 11.

Carrier aggregation is also enhanced further in Release 11. The enhancements include improved support for different downlink/uplink configurations in the TDD mode on different component carriers that are on different frequency bands. Different timing advances on different component carriers is also supported.

The FeICIC feature includes support for an interference cancellation receiver at the UE. The network may configure the UE with cell-specific reference signal (CRS) assistance information of the interfering cells in order to aid the UE to mitigate the interference from CRS of those cells.

MBMS service continuity includes support for multicarrier MBMS deployment. The network can inform the UE about the carrier frequencies used and which services are available on each frequency. The UE can also indicate the network on which MBMS services is interested, and the network can use this information in its mobility management so that the UE in question is kept on the appropriate carrier frequency so that it can continue to receive the selected services.

Release 11 also contains a long list of various SON enhancements. These are discussed together with earlier SON features in Chapter 11.

Mobile relays study item was also part of Release 11, but due to lack of time this work item was postponed to Release 12.

The LTE-A development work is continuing, and each new release will bring further improvements. The time span between releases is currently about 18 months.

Release 12 LTE-A enhancements are discussed in Chapter 12, Future Trends. However, at the time of this writing, Release 12 work is still ongoing, and therefore it is not quite sure yet what it will contain.

REFERENCES

[1] 3GPP TR 25.913, v 8.0.0, *Requirements for Evolved UTRA (E-UTRA) and Evolved UTRAN (E-UTRAN)*, 2008.

[2] 3GPP TS 36.211, v 8.9.0, *Evolved Universal Terrestrial Radio Access (E-UTRA); Physical Channels and Modulation*, 2009.

[3] Rec ITU-R M.2012.

CHAPTER 4

OFDMA

4.1 Introduction

Orthogonal frequency division multiplexing (OFDM) as a technology is quite old; 3GPP by no means invented it. It has been used earlier for many systems such as wireless LAN standards 802.11a/g/n, DVB-T, DVB-H, and many others. It is also been used in IEEE 802.16m (mobile WiMAX) [1], which is the other 4G standard in existence today. OFDM is also used in many cable systems. Therefore, it is a proven technology.

Orthogonal frequency division multiple access (OFDMA) is simply a variant of OFDM whereby different OFDM subcarriers can be allocated to different users. This transforms OFDM into multiple access scheme, enabling several users to access system resources simultaneously.

From the user point view the different air interface technology means that a new handset is required before moving to a 4G network. In practice all 4G capable handsets will be at least dual-mode (3G/4G) phones, or even 2G/3G/4G phones (though in Japan the 2G network is already closed, so the new Japanese mobile phones are typically dual-mode 3G/4G handsets).

OFDM has many qualities that make it especially suitable for wideband communications. It can make use of wide carriers very well. This is important in 4G since higher data rates require more spectrum and wider carriers. In systems that are based on CDMA (or TDMA, for that matter) the problem with wide channels is that such channels become very frequency selective (i.e., the signal gets distorted when transmitted over the air), and the distortion is dependent on the frequency used (see Figure 4.1). Designing a receiver for frequency selective channel is difficult, especially if the receiver is in a mobile device that has limited amount of processing capacity and power in its use. The complexity of such receivers in wideband channels makes it impractical to use WCDMA with, say, a single 20-MHz carrier.

OFDM divides the wide carrier into numerous narrowband subcarriers and handles them (almost) individually. These subcarriers are so narrow that the channel response is frequency nonselective (frequency-flat) (i.e., the receiver can assume that the received narrowband signal is equally distorted over the whole frequency area of the subcarrier). In other words, the channel bandwidth of a frequency flat channel should be smaller than the coherence bandwidth. For such a signal, it is much easier to design an efficient

48   OFDMA

FIGURE 4.1
Frequency selective fading.

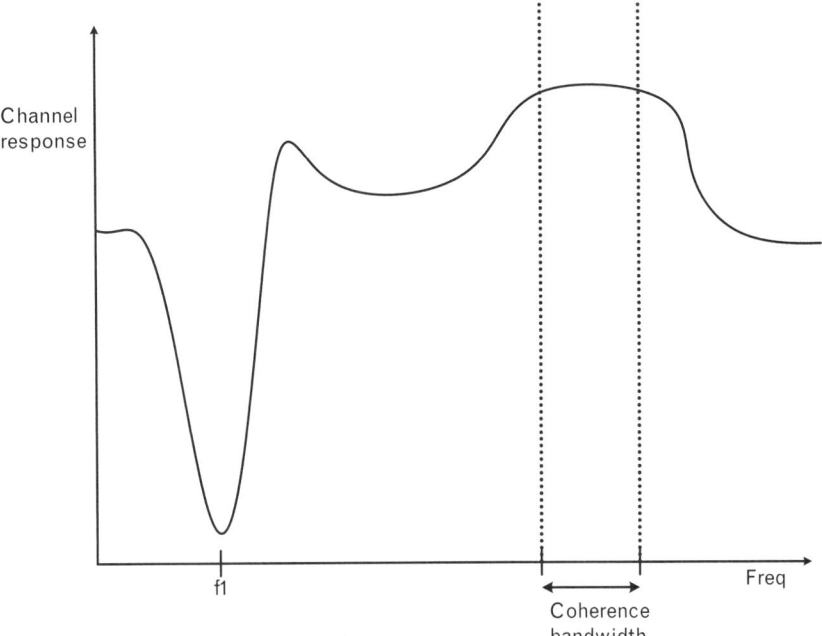

receiver. However, note that in practice some parts of the frequency channel may remain unusable even if the subcarrier bandwidth is smaller than the coherence bandwidth. For example, in Figure 4.1 there is a very deep fade around frequency f1. Therefore this frequency channel should be left unused. Fortunately, in LTE it is possible to use channel response (i.e., channel quality measurements) as one scheduling parameter in the eNodeB scheduler: the objective is to schedule for each UE only those frequency channels that offer the best channel response for each particular UE.

The subcarriers are packed very close to each other and normally these carriers would interfere with each other. However, the "O" in OFDM stands for *orthogonal*, and that is how the arrangement works. The subcarrier frequencies are chosen so that the subcarriers are orthogonal with each other, and as a result the subcarriers do not interference with each other. There is no need to have a guard band between subcarriers. Figure 4.2 depicts three orthogonal subcarriers; frequency f1 has the peak amplitude for the first subcarrier, but subcarriers 2 and 3 have zero amplitude on that frequency. Similarly, f2 provides the peak amplitude for the second subcarrier, but zeros for subcarriers 1 and 3.

The orthogonality requires that the subcarrier spacing is $\Delta f = \dfrac{k}{T_U}$, where T_U is the useful symbol duration in seconds (the receiver side window size), and k is a positive integer, usually 1. In LTE the symbol duration is 66.67 us and the subcarrier spacing is thus 15 kHz [2]. Note that as a result of subcarrier orthgonality, OFDM is very spectral efficient because

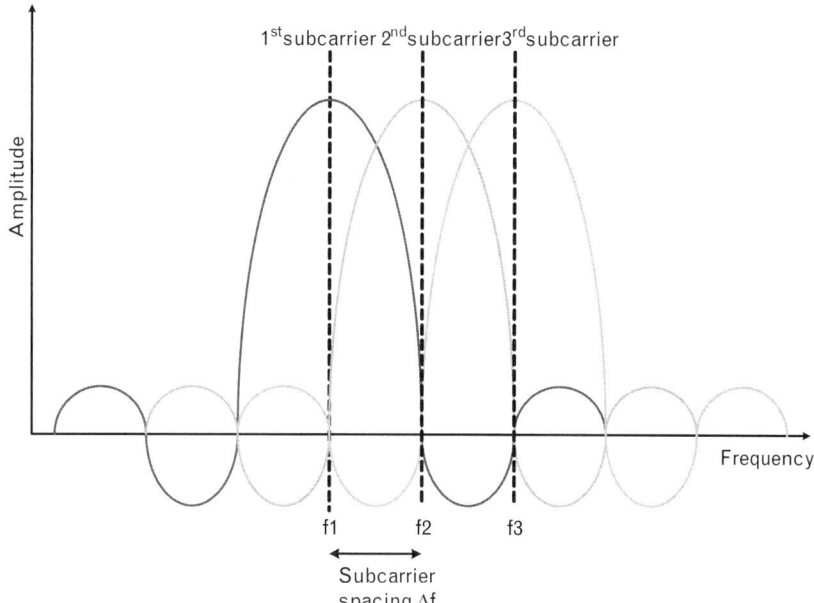

FIGURE 4.2
Orthogonal subcarriers.

the subcarriers can be packed tightly. On the other hand, OFDM is very sensitive to frequency offset. If frequency synchronization is lost, then also the subcarrier orthogonality is gone, and the receiver will suffer from serious intercarrier interference (ICI). Figure 4.3 shows a system where the orthogonality is lost, resulting in ICI. In this example, the second subcarrier has drifted slightly toward the first subcarrier, and this has resulted in serious ICI, affecting both subcarriers.

A typical radio channel also suffers from multipath effects. The receiver may get delayed versions of the same signal, in which case the delayed symbol received from a long propagation path may overlap with the next symbol at the receiver resulting in intersymbol interference (ISI) (see Figure 4.4). In modern systems, the data rates are ever higher, and consequently the symbol lengths become shorter. As a result it could be that ISI is not the result of only two successive symbols interfering with each other, but the ISI may span over several symbol periods. Designing a receiver for such a system is very challenging. However, OFDM can suppress ISI quite efficiently. OFDM employs a large number of narrowband carriers, which are modulated at a low symbol rate. This results in a long symbol duration which limits the effects of ISI and makes it feasible to use cyclic prefixes to remove ISI altogether.

A cyclic prefix acts as a guard interval, and it is added between each OFDM symbol. Typically it consists of a copy of the end part of the same OFDM symbol. It is added in the beginning of the symbol at the transmitter and removed at the receiver, all while removing possible ISI effects (see Figure 4.5). The guard interval should be longer than the length of the

50 OFDMA

FIGURE 4.3
Subcarriers with lost orthogonality.

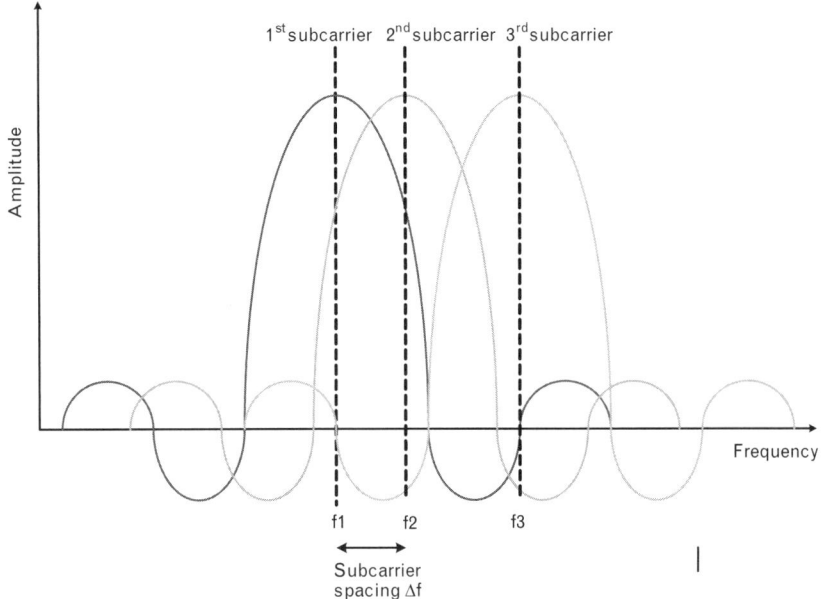

FIGURE 4.4
Intersymbol interference.

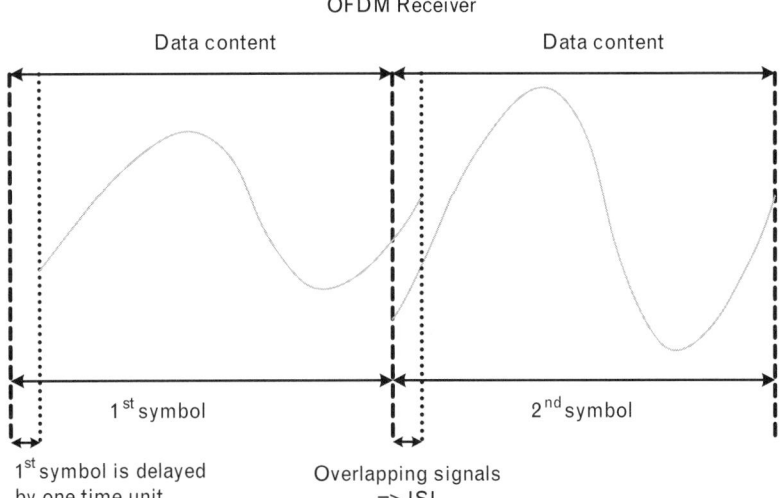

longest ISI period expected. Note that the guard interval should not be longer than necessary, since too long a guard interval is channel capacity wasted, after all.

The guard interval could also be empty if we were interested only in the ISI removal. However, by using a cyclic prefix to fill the guard interval, the resulting signal is cyclically extended. Such a signal maintains the subcarrier orthogonality, which would be lost with empty guard intervals, and,

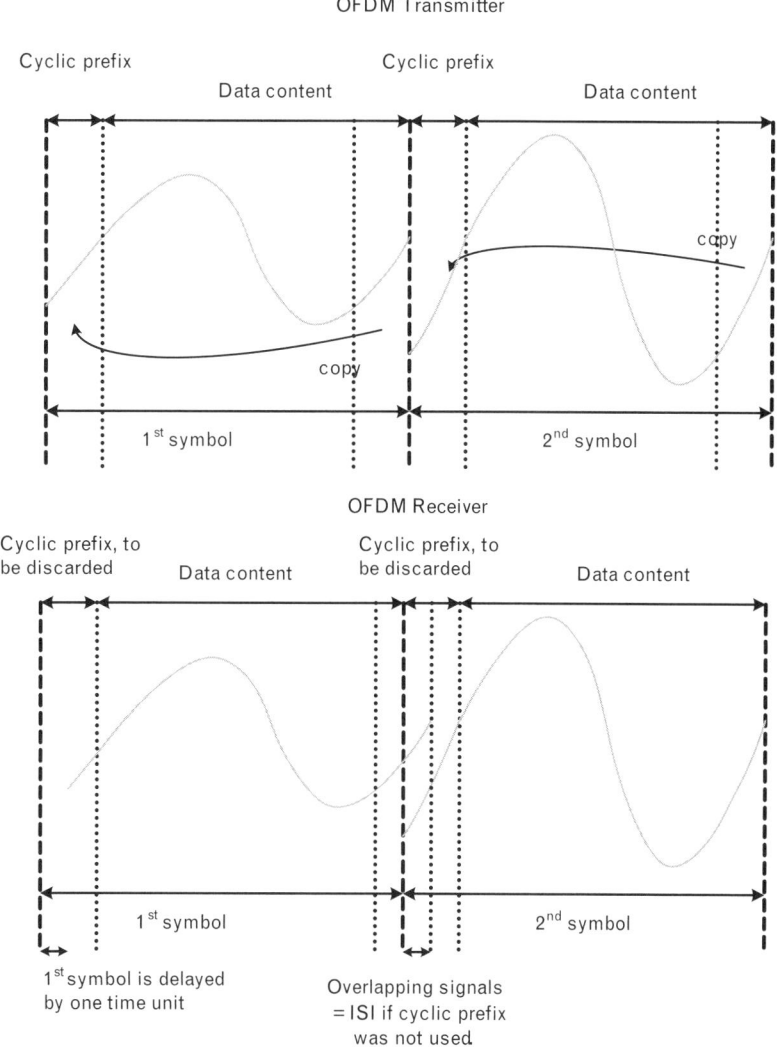

FIGURE 4.5
Intersymbol interference suppression with cyclic prefix.

as we discussed earlier in this chapter, lost subcarrier orthogonality would result in ICI.

4.2 OFDM Principles

This section discusses the principles of OFDMA as it is defined and used in LTE. OFDMA as such is also used in LTE-A—the basic principles are the same in both cases. However, LTE-A contains other technical enhancements that will improve the capabilities of the system. These are briefly discussed in the end of the section.

52 OFDMA

OFDM is used differently in LTE downlink and uplink. We will first concentrate on the downlink, and the uplink is discussed in Section 4.3. Also, if not mentioned otherwise, the text here refers to the FDD mode. The OFDM principles are the same for the TDD mode, too, but the physical layer is inevitably somewhat different in FDD and TDD modes.

Figure 4.6 presents a simplified OFDM signal processing chain. Note that this is not a description of the whole LTE layer 1—that is a much more complex entity. This section describes only the OFDM principles; the LTE physical layer in general is then discussed in Chapter 5.

Since the OFDM scheme transmits the data via several narrowband subcarriers, the data bits have to first undergo a serial-to-parallel conversion, one parallel stream for each subcarrier. Next an inverse-FFT (IFFT) algorithm transforms the modulated subcarriers from frequency domain to time domain samples. Fast Fourier transform (FFT) [3] and its inverse, IFFT, are widely used algorithms in engineering. FFT transforms the signal representation from the time domain to the frequency domain, and IFFT does the opposite. In case of OFDM, IFFT is used at the transmitter and FFT at the receiver.

After the IFFT function, the cyclic prefix is added. In the LTE physical layer, all timing is derived from the basic time unit Ts = 1/30720000 seconds. The radio frame is then:

$$T_{frame} = 307200 \star T_s = 10 \text{ ms}$$

Each radio frame is divided into 10 subframes, and each subframe further divided into two slots. A subframe is 1 ms in duration and a slot is 0.5 ms [2, 4]. A subframe is the smallest time unit that can be scheduled by the network scheduler (see Figure 4.7).

Futhermore, each slot is divided into symbols, either six or seven symbols per slot. In the "normal" mode the slot contains seven symbols, and in the extended mode there are six symbols in every slot. In both cases

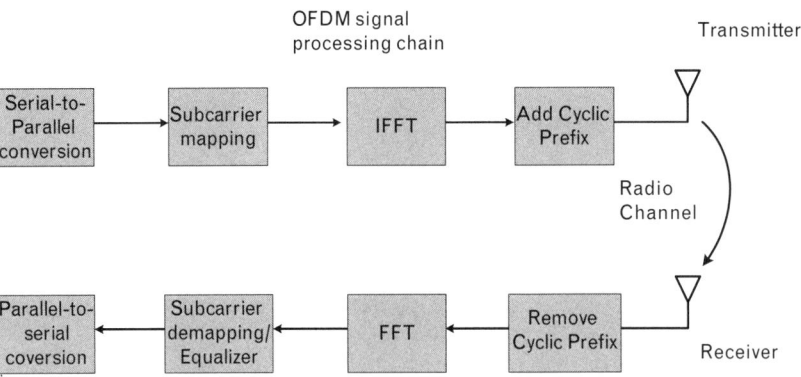

FIGURE 4.6
Simplified OFDM signal processing chain.

4.2 OFDM PRINCIPLES

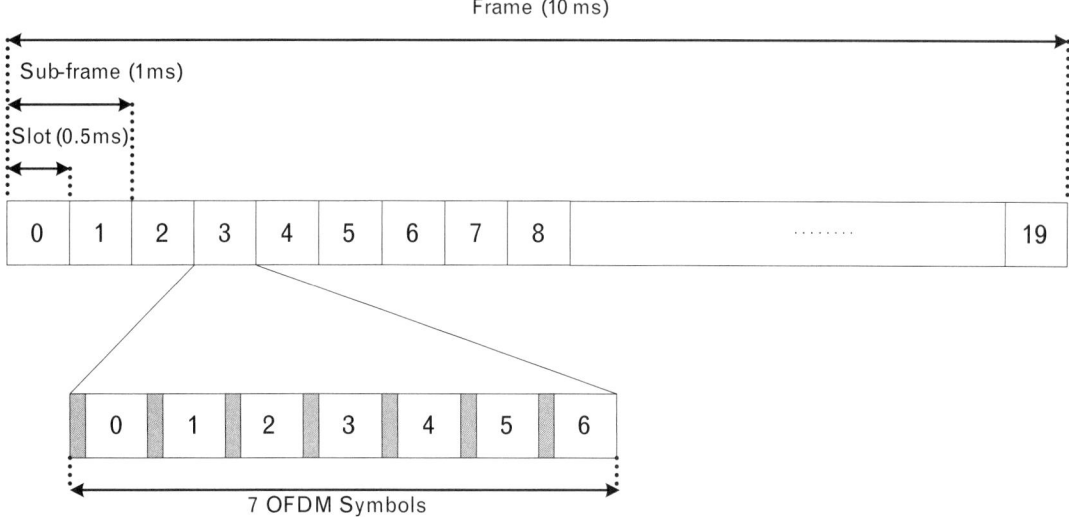

FIGURE 4.7 *FDD mode radio frame structure.*

the effective symbol length Tu is the same, $Tu = 2048 \star Ts = 66.7$ us. The cyclic prefix is added after every symbol, and for the first symbol in each slot (number #0) the cyclic prefix is $Tcp = 160 \star Ts = 5.2$ us, while for the others (numbers 1–6) the length is $Tcp = 144 \star Ts = 4.7$ us (see Figure 4.8).

FIGURE 4.8
LTE OFDM symbols.

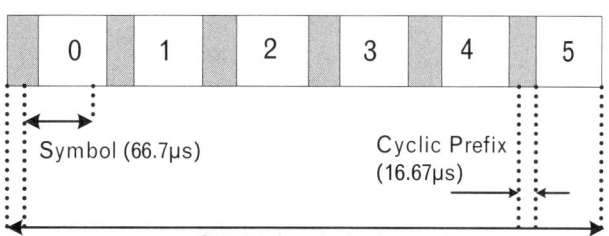

These values are for the normal mode. In the extended mode the cyclic prefix is longer: $Tcp\text{-}e = 512 * Ts = 16.7$ us. The extended mode is to be used in situations where a large signal delay spread is expected. The mode to be used is signaled by the network. Note that the normal cyclic prefix can handle ISI resulting from delay spreads up to 1.4 km (speed of signal[1] × 4.69 μs = 1406m) difference in signal propagation paths. The extended mode cyclic prefix is much longer and can remove ISI effects until the propagation path length difference is 5 km (speed of signal x 16.7 μs = 5007m) [2, 4].

The cyclic prefix takes about 7.5 percent of the channel capacity in the normal mode. This is a significant slice of the overall capacity available, but it should not be considered as wasted capacity since the signal quality is much better if ISI does not cause problems, and it may be possible to employ higher order modulation schemes as a result.

In the OFDM air interface the number of subcarriers can vary between 128 (with a 1.4-MHz frequency band) and 2048 (20-MHz band). The spacing between subcarriers is 15 kHz. Not all subcarriers are used for data (e.g., the band's "center" carrier, direct current (DC), is not used for anything and neither are the guard carriers on either side of the frequency band). The DC carrier is intentionally left empty because from the empty subcarrier the UE can identify the center of the frequency band. The guard carriers are used to minimize interference from adjacent frequency bands.

In the air interface the data to be transmitted is mapped into resource elements and resource blocks (see Figure 4.9). A resource block is a time-frequency entity, consisting of one slot in the time domain and 12 subcarriers in the frequency domain. In other words, a resource block is a 0.5 ms * 180 kHz slice of the time-frequency space. The number of resource blocks on LTE bands vary from 6 (1.4-MHz band) to 100 (20-MHz band). The set of resource blocks over the whole band is called the resource grid. If several antenna ports are used (i.e., if MIMO is employed), then each antenna port has its own resource grid.

A resource block is built of resource elements. One resource element consists of one symbol on one frequency subcarrier. A resource element is the smallest resource unit that can be addressed and allocated in the air interface, though when it comes to scheduling resources to users, the unit used is a resource block. All together, a resource block contains 12 * 7 = 84

TABLE 4.1 RESOURCE BLOCK PARAMETERS

CONFIGURATION	SUBCARRIER BW	NUMBER OF SUBCARRIERS	NUMBER OF OFDM SYMBOLS
Normal cyclic prefix	15kHz	12	7
Extended cyclic prefix	15kHz		6
	7.5kHz	24	3

1. Speed of signal = speed of light = 299 792 458 meters per second.

FIGURE 4.9
Resource block and resource element.

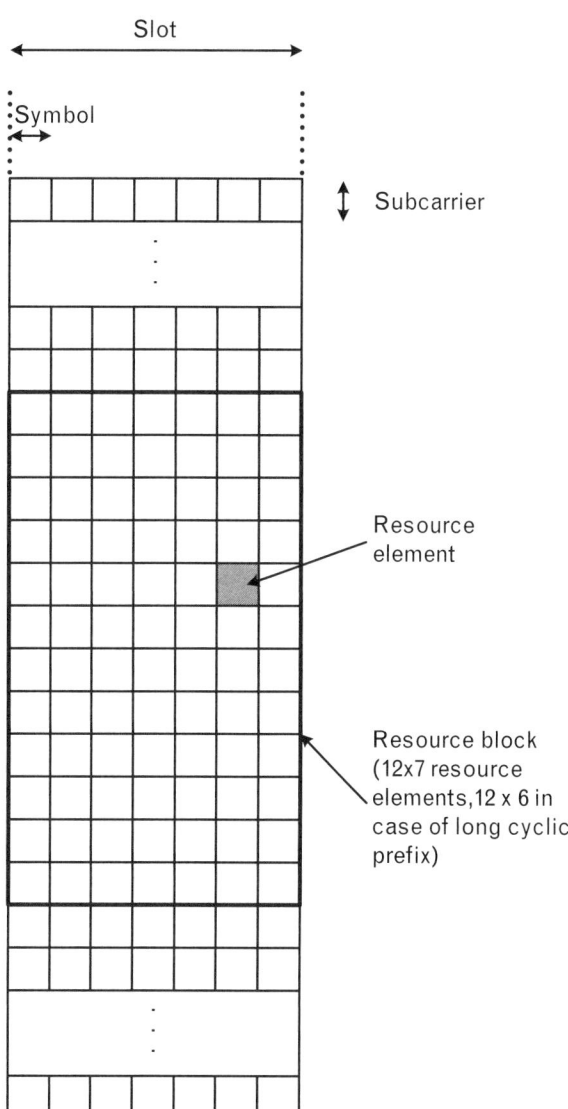

resource elements (or 12 * 6 = 72 resource elements in case the extended cyclic prefix is employed) [2].

The air interface resource scheduling is controlled by the MAC layer in eNodeB. Both uplink and downlink shared channels have their own schedulers. The smallest scheduling time unit for user services is one TTI, which is two successive resource blocks (i.e., 12 subcarriers * 1 subframe). The actual scheduling algorithm itself is not standardized; it is up to eNodeB vendor to implement as efficient a scheduler as possible. However, its working principle is that it will collect information on channel conditions and data transmission requests, and, based on this information, it assigns suitable subcarriers for each user so that the overall system throughput is maximized.

4.3 LTE Uplink—SC-FDMA

It can also modify other parameters such as the channel modulation and coding rate separately for each resource block. More about the scheduling can be found in Chapter 6, MAC Layer.

One problem with OFDM, as used in the LTE downlink, is its high peak to average power ratio (PAPR). In theory at least it is possible that all subcarriers use the maximum power level simultaneously, and as a result the combined carrier can have a very high power level. This can be handled at the base station transmitter (i.e., on the downlink), but for user equipment (UE), a high PAPR means inefficient power amplifier performance and thus short battery life. Therefore for the LTE uplink, another technique called single-carrier FDMA (SC-FDMA) has been chosen.

SC-FDMA is a hybrid scheme that combines the low PAPR of a single carrier system with many advantages of an OFDM system. Figure 4.10 shows the high-level block diagram of the SC-FDMA processing chain. As one can see, it is very similar to OFDM, except that in the beginning of the transmission chain the time-based signal is converted to a frequency-based signal using FFT. This has the effect of spreading the information bits onto all subcarriers, instead of the downlink OFDM where bits are sent in parallel, in separate subcarriers. As a consequence of this spreading, the PAPR experienced will be much smaller. Those subcarriers not used for uplink transmission will be set to zero. The FFT function is then followed by much of the same processing chain as with OFDM. IFFT is used to convert the signal to time domain, and after that cyclic prefixes are added in a similar way as in OFDM. The three resource concepts (resource element, resource block, and resource grid) are defined in exactly the same way as in OFDM. Also the subcarrier spacing is 15 kHz in both in OFDM and in SC-FDMA.

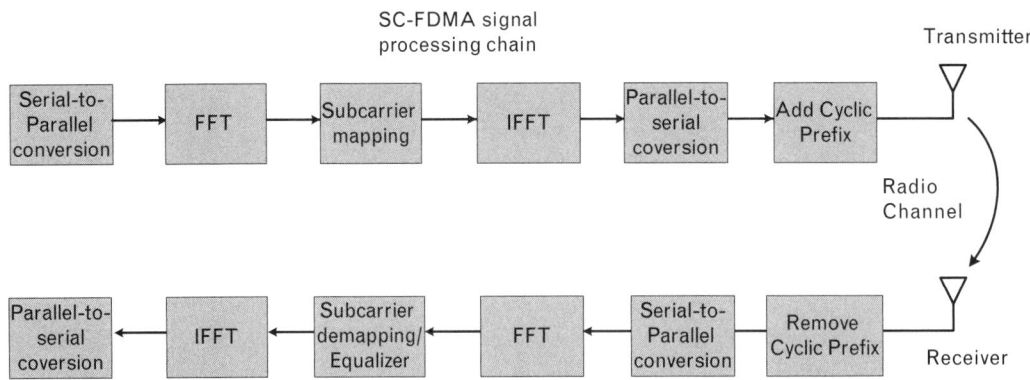

Figure 4.10 *SC-FDMA signal processing chain.*

The SC-FDMA receiver block diagram is a mirror image of the transmitter diagram.

In addition to saving power, SC-FDMA has some other advantageous features:

- Since data bits are spread over wide channels, it offers some level of frequency diversity.
- SC-FDMA is not as sensitive to frequency offset as OFDM.
- Lower complexity transmitter (important for UEs).

4.4 Summary of OFDMA

As seen in this chapter, OFDM has many good qualities and that is why it was chosen as the air interface technology for LTE. The list of advantages includes the following issues, among others:

1. Using orthogonal subcarriers means that the subcarriers can be packed tightly, resulting in high spectral efficiency.
2. Orthogonality also means no intercarrier interference, and no need for intercarrier guard bands.
3. Long symbol time and guard intervals remove intersymbol interference.
4. OFDM is flexible on spectrum bands: it can exploit both relatively narrow and very wide bands.
5. The large number of subcarriers enables the use of efficient scheduling algorithms, whereby each user is allocated the best nonfading subcarriers/resource blocks from the time/frequency space.
6. Implementing OFDM is easy with FFT/IFFT algorithms.
7. Suitable for use when implementing single-carrier networks (for services such as MBMS).

However, it was also pointed out that the basic OFDM has some disadvantages. The two most important ones are:

1. Sensitive to frequency offset;
2. High peak-to-average power.

The first problem may cause the loss of subcarrier orthogonality, which in turn will introduce lots of interference from adjacent subcarriers (ICI). To overcome this, OFDM transmission contains two types of signals that can be used for frequency estimation: synchronization signals and cell-specific reference signals [2, 5]. These are discussed in Section 5.5.1.

High PAPR, which was discussed in Section 4.3, would be a serious problem for UE transmitters, and therefore LTE uplink uses a modified form of OFDM called SC-FDMA.

In the next chapter we will continue with the LTE air intrface.

REFERENCES

[1] IEEE 802.16m Task Group (Mobile WirelessMAN), http://ieee802.org.

[2] 3GPP TS 36.211, v 11.4.0, Evolved Universal Terrestrial Radio Access (E-UTRA); Physical Channels and Modulation, 09/2013.

[3] Brigham, E. O., *The Fast Fourier Transform,* New York: Prentice-Hall, 2002.

[4] 3GPP TS 36.300, v 11.7.0, Evolved Universal Terrestrial Radio Access (E-UTRA) and Evolved Universal Terrestrial Radio Access Network (E-UTRAN); Overall Description; Stage 2; 09/2013.

[5] Wang, Oi, Christian Mehlfuhrer, and Markus Rupp, "Carrier Frequency Synchronization in the Downlink of 3GPP LTE," *The 21st Annual IEEE International Symposium on Personal, Indoor and Mobile Radio Communications (PIMRC'10),* Istanbul, Turkey, pp. 939–944.

CHAPTER 5

Air Interface: Physical Layer

5.1 Introduction

In this chapter we will continue to handle the air interface. The previous chapter already presented OFDM and SC-FDMA basics, and now we are looking at other air interface details: starting with the physical layer and in the following chapters continuing with protocol layers, protocols, and the typical procedures that can take place in this interface.

The air interface is well specified—every single detail in it has to be agreed and put into standards because this interface is truly open. There are many radio access network equipment vendors and numerous user equipment manufacturers, and all of these devices have to work seamlessly together over the air interface.

LTE follows a layered protocol model as shown in Figure 5.1, with defined interfaces between layers. However, it is good to remember that these interlayer interfaces are internal interfaces for UEs and eNodeBs; they are not open interfaces such as the air interface. They are not standardized, and the manufacturer may implement them in the most appropriate way. Also the functionality in protocol layers may be customized (e.g., sometimes it may be more efficient to move some functionality to another protocol layer).

There are three separate channel concepts in LTE: physical, transport, and logical channels (see Figure 5.2).

Logical channels define what type of data is transferred. These channels define the data transfer services offered by the MAC layer; that is, the concept of logical channels is used in the interface above the MAC.

Transport channels define how and with which type of characteristics the data is transferred by the physical layer. These channels are used in the interface between the MAC and the physical layers.

Physical channels define the exact physical characteristics of the radio channels. These are the channels used below the physical layer; that is, in the air interface.

60   AIR INTERFACE: PHYSICAL LAYER

FIGURE 5.1
Air interface protocol stack.

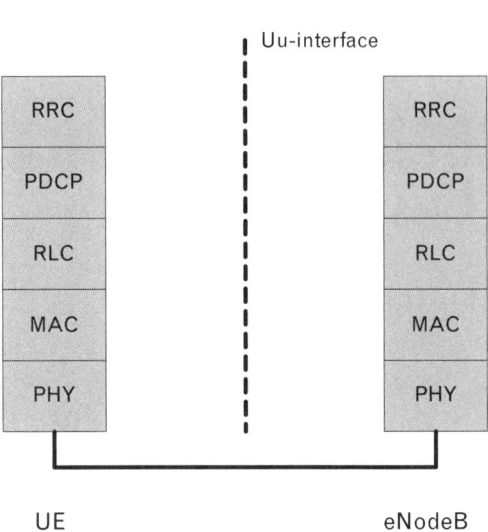

FIGURE 5.2
Air interface channel concepts.

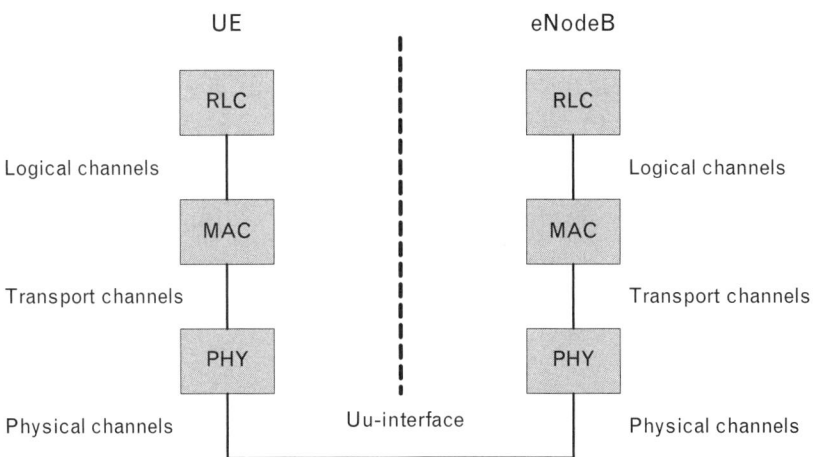

5.2   Physical Layer—General

The lowest layer in the air interface is the physical layer. It is the first entity facing the air interface in the user terminal and in the eNodeB, transmitting and receiving bit streams. Therefore, its functions are closely coupled with the air interface timing, indicating that its processes are usually tightly time constrained. That is, the physical layer must react on inputs and perform its tasks very quickly. This is also the reason why the physical layer is typically implemented in hardware, as opposed to software running on a CPU, as hardware implementations are much quicker.

As discussed in Chapter 4, the air interface technologies in the downlink and uplink are slightly different. Also, the physical layer entities in the UE and in the eNodeB are different, especially implementationwise. The eNodeB can use more computational-intensive solutions, more complex receiver algorithms, bigger processors, and—especially important—there is no shortage of electric power. The UE, on the other hand, is a mass-produced device where each component has gone through a strict cost-performance analysis (i.e., its components can fulfill the task they are designed to do but no more). These handheld devices are typically battery powered; therefore, energy saving is essential.

The air interface frame structure for the FDD mode was briefly discussed in Chapter 4. However, since the TDD mode has a different structure, both frame types are discussed next.

In the FDD mode each radio frame (10 ms) is divided into 10 subframes, and each subframe further divided into two slots [1]. A subframe is 1 ms in duration and a slot is 0.5 ms. A subframe is the smallest unit that can be scheduled by the network scheduler. The FDD mode radio frame is also known as the type 1 frame (see Figure 5.3). The FDD mode can be either half-duplex or full-duplex, depending on UE capabilities. A UE that can transmit and receive simultaneously is a full-duplex UE. If it can do only one of them at a time, then the device is a half-duplex UE. Note that in the FDD mode, the uplink and the downlink use different frequencies.

In the TDD mode, a type 2 frame is used. Each 10 ms radio frame consists of two 5-ms half-frames. Moreover, each half-frame consists of eight 0.5-ms slots. In addition, a half-frame has three special fields in a special subframe: downlink pilot time slot (DwPTS), guard period (GP), and uplink pilot time slot (UpPTS). The total length of DwPTS, GP, and UpPTS fields together is 1 ms (see Figure 5.4).

Both 5-ms and 10-ms switch-point periodicities are supported. The switch-point periodically indicates how often the transmission switches between the uplink and the downlink. Note that even though the TDD mode is inherently suitable for asymmetric data connections, in LTE there are only seven different uplink-downlink configurations for the eNodeB to

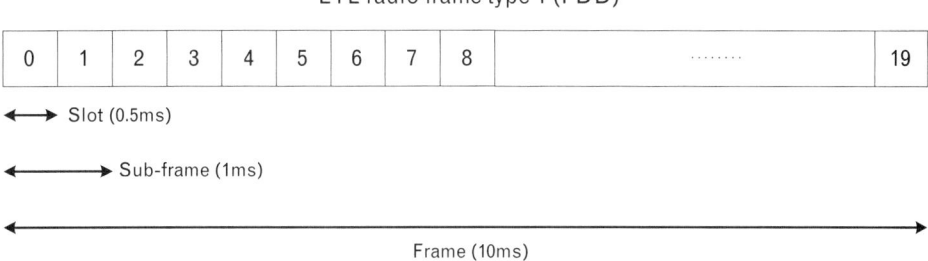

FIGURE 5.3 *FDD mode radio frame structure, type 1 frame.*

62 AIR INTERFACE: PHYSICAL LAYER

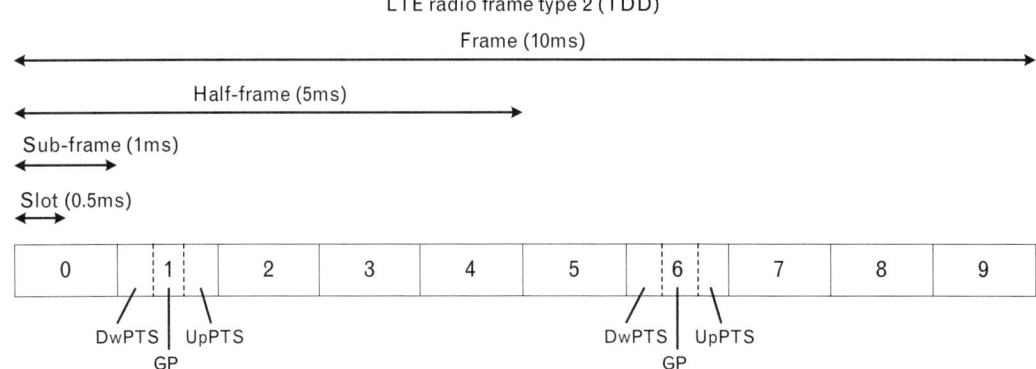

FIGURE 5.4 *TDD mode radio frame structure, type 2 frame.*

choose from. Table 5.1 shows the various uplink-downlink combinations for both periodicity values.

Subframe 1 is always a special subframe (i.e., the switching point), and in addition subframe 6 is also a special subframe when 5 ms switch-point periodicity is used. Subframes 0 and 5 are always reserved for the downlink, and subframe 2 for the uplink.

5.3 Physical-Layer Processing

The downlink physical-layer processing of data received from higher layers includes the following steps [2]:

- CRC insertion;
- Channel coding;
- Physical-layer hybrid-ARQ processing;

TABLE 5.1 UPLINK-DOWNLINK ALLOCATIONS FOR THE TDD MODE

Configuration Index	Switch-Point Period (ms)	Subframe Number										
		0	1	2	3	4	5	6	7	8	9	
0	5	D	S	U	U	U	D	S	U	U	U	
1	5	D	S	U	U	D	D	S	U	U	D	D = downlink
2	5	D	S	U	D	D	D	S	U	D	D	U = uplink
3	10	D	S	U	U	U	D	D	D	D	D	S = switch-point
4	10	D	S	U	U	D	D	D	D	D	D	
5	10	D	S	U	D	D	D	D	D	D	D	
6	5	D	S	U	U	U	D	S	U	U	D	

5.3 PHYSICAL-LAYER PROCESSING

- Rate matching and channel interleaving;
- Scrambling;
- Modulation;
- Layer mapping and precoding;
- Mapping to assigned resources and antenna ports.

See Figure 5.5 for the physical layer downlink data path. These steps are discussed in detail in the following sections. The data processing in the uplink direction follows the same principles and will not be discussed separately.

5.3.1 CRC Insertion

Physical downlink shared channel (PDSCH) is the main physical channel for carrying user data in the downlink direction. The physical layer gets the data in transport blocks from the MAC layer, at a rate of maximum of two data blocks per TTI. The first task for the physical layer is to calculate and attach cyclic redundancy check (CRC) bits to the block. The CRC is a form of a block code for error checking at the receiver.

A block code manipulates the data one block at a time. The encoder adds some redundant bits to the block of bits, and the decoder uses them to determine whether an error has occurred during the transmission. If an error is found, the receiving entity can ask for a retransmission of the same data. The CRC bits are then discarded at the receiver. The more CRC bits there are, the better the code is to detect errors. But the added redundancy also consumes bandwidth, so a good balance must be found.

In the LTE PDSCH, the length of CRC is 24 bits per block. The other channels in LTE may also use CRC lengths of 16 and 8 bits. Even the 24-bit CRC in LTE has two versions, A and B, depending on what kind of generator polynomial is used to calculate the CRC parity bits. The polynomials used in LTE are given in Table 5.2. The PDSCH uses $g_{CRC24A}(D)$ polynomial.

For example, the CRC generator polynomial $D8 + D7 + D4 + D3 + D + 1$ means that the polynomial bit string is 110011011. The data block is divided modulo 2 by the generator polynomial, and the remainder becomes

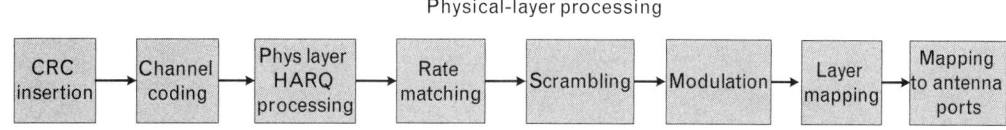

FIGURE 5.5 *Physical layer downlink data path.*

Table 5.2 CRC Generator Polynomials

CRC Algorithm	Generator Polynomial
24A	$D^{24} + D^{23} + D^{18} + D^{17} + D^{14} + D^{11} + D^{10} + D^7 + D^6 + D^5 + D^4 + D^3 + D + 1$
24B	$D^{24} + D^{23} + D^6 + D^5 + D + 1$
16	$D^{16} + D^{12} + D^5 + 1$
8	$D^8 + D^7 + D^4 + D^3 + D + 1$

the checksum field. The data block size can be between 40 and 6,144 bits. If there are less than 40 bits to send, then the data block has to be padded with filler bits. If there are more bits to send than the maximum code block size (6,144 bits), then the bit sequence is segmented into smaller data blocks and an additional sequence of CRC bits is attached to each block.

Because CRC and other block codes are typically used with retransmission schemes, they cannot be used with channels that have very tight timing constraints because there is no time to wait for a retransmission. CRC insertion is specified in [3].

5.3.2 Channel Coding

In addition to error checking performed by the block coding function, the LTE employs two forward error correction (FEC) schemes: convolutional codes and turbo codes [3]. The difference between a block code and a FEC function is that whereas the block code simply indicates to the receiver whether or not the data block was received correctly, FEC codes try to fix the errors found, up to a certain point. FEC codes achieve this capability by adding redundancy to transmitted data. The ratio between the number of input bits and the number of output bits from the coding function is called the coding rate. In LTE, most channels use coding rate of 1/3. That is, after the channel coding, there are three times more bits to be transmitted. Therefore, channel coding should not be used unnecessarily, and a suitable code rate should be carefully considered.

Both the turbo coding and the convolutional coding that are used in LTE produce so called systematic code, as opposed to nonsystematic code. In a systematic code, all redundant (or parity) bits are added to the end of the codeword, whereas in nonsystematic code the redundant bits are mixed with information bits at the coding function output. Both in the turbo coding and in the convolutional coding, the output has three streams: the first stream contains the information bits, and the two other streams have parity bits. However, as explained later, these streams will be combined after channel coding in the rate matching function.

5.3.2.1 Turbo Codes

Turbo codes are a relatively new invention, first discussed in 1993 [4]. They are found to be very efficient, because they can perform close to the theoretical limit set by the Shannon Law [5].

In turbo coding, the output of the decoding process is used to readjust the input data. This iterative process improves the quality of the decoder output, although the returns from the process diminish with every iterative loop. The turbo encoder specified in the LTE is shown in Figure 5.6. This encoder is a parallel concatenated convolutional code (PCCC). It consists of two convolutional encoders in parallel separated by an interleaver. The encoders are recursive and systematic. The task of the interleaver is to randomize the data before it enters the second encoder. The interleaver consists of a rectangular matrix. It performs both intra-row and inter-row permutations for the input bits, and the output bit sequence is pruned by deleting these bits, which were not part of the input bit sequence. The last phase of turbo encoding is called Trellis termination. It is performed by taking the tail bits from the shift register feedback after all information bits are encoded. Tail bits are padded after the encoding of information bits. The output of the Trellis termination stage is three bit streams.

The turbo decoder is depicted in Figure 5.7. It consists of two soft input soft output (SISO) decoders connected by interleavers and a de-interleaver. The extrinsic information is relayed back from the output of the second decoder to the input of the first decoder. Each iteration improves the estimate of the extrinsic information, which again improves the estimate for the de-

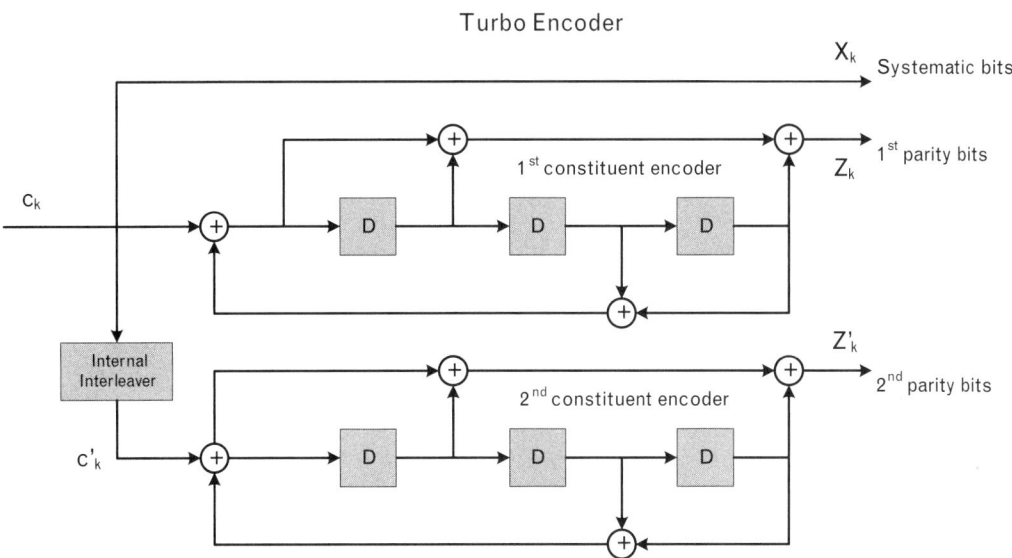

FIGURE 5.6 *Turbo encoder.*

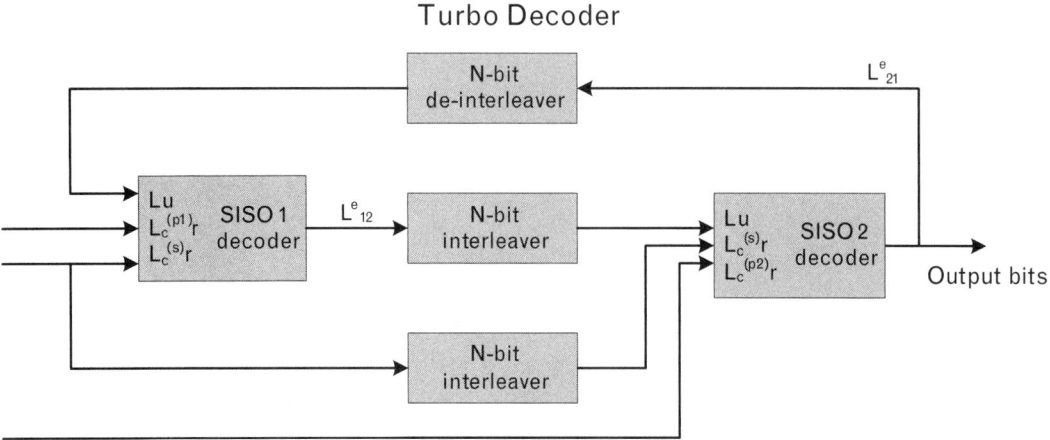

FIGURE 5.7 *Turbo decoder.*

coded data. The data is processed in the iterative loop on a block-by-block basis.

5.3.2.2 Convolutional Codes

Convolutional codes are different from block codes in that they operate continuously on streams of data. They also have a memory, which means that the output bits do not depend only on the current input bits, but also on several preceeding input bits. A convolutional code can therefore be described using the format (n, k, m), where n is the number of output bits per data word, k is the number of input bits, and m is the length of the coder memory. The code rate of the convolutional code is thus similar to block codes:

$$Rc = k/n \tag{5.1}$$

The convolutional coder adopted by LTE is shown in Figure 5.8. It is a combination of shift registers (D) and exclusive-OR functional units. In the end of the data sequence to be encoded, the convolutional coder adds m-1 zeros to the output sequence. This is done periodically to force the encoder back to its initial state. This technique improves the error protection for the encoded bits at the end of the message. The structure of the encoder

5.3 PHYSICAL-LAYER PROCESSING

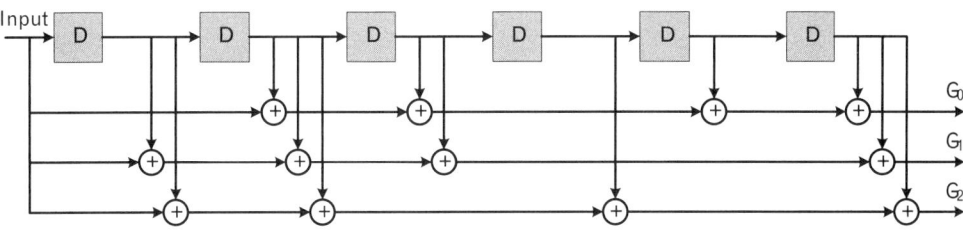

Generator polynomials:
$G_0 = 133$ (octal), for the first parity stream
$G_1 = 171$ (octal), for the second parity stream
$G_2 = 165$ (octal), for the third parity stream

FIGURE 5.8 *Convolutional 1/3 rate encoder.*

is rather simple and its operation straightforward. The decoder, however, is something completely different.

The optimal convolutional decoder is the maximum likelihood sequence estimator (MLSE) [6], which is based on the idea that for a finite sequence of bits, the receiver generates all possible sequences the encoder could possibly have sent. Next, the receiver compares the actual received bit sequence with each of the generated sequences and calculates the Hamming distance for each pair. The minimum Hamming distance should identify the most likely transmitted sequence.

The MLSE method provides the most efficient convolutional code decoder, but the problem is that once the size of the transmitted bit sequences increases, the complexity of the MLSE algorithm becomes unmanageable. A solution to this problem is to use the Viterbi algorithm, which estimates the MLSE algorithm well enough still to be efficient.

The actual theories behind the MLSE and Viterbi algorithms are outside the scope of this book. However, convolutional codes and maximum likelihood decoders are explained well in [7].

Convolutional decoders can be either hard or soft decision decoders. This division refers to the way the decoders receive the bit information from the demodulator. In the hard decision method, the demodulator output is either a 0 or 1. In the soft decision method, the demodulator returns not only the received bit (0 or 1), but also an estimation of the reliability of this decision. The estimation can be based on such things as the current received signal level. The decoder can then use the estimation data as one parameter in its maximum likelihood decoding algorithm.

Convolutional decoders work well against random errors, but they are quite vulnerable to bursts of errors, which are typical in mobile radio systems. This problem can be eased with interleaving, which spreads the erroneous bits over a longer period of time and thus makes the convolutional decoder more efficient.

5.3.2.3 Summary

The channel coding schemes used in LTE are given in Tables 5.3 and 5.4. Note that on some physical channels in LTE, both FEC and block codes are used. First the channel decoder (either a convolutional or turbo decoder) tries to correct as many errors as possible, and then the block decoder (CRC check) gives its judgment on whether the resulting information is good enough to be used in the higher layers.

Convolutional coding can be used for low data rates, and turbo coding for higher rates. At higher bit rates, turbo coding is more efficient than convolutional coding. Turbo coding is not suitable for low rates, as it does not perform well on short blocks of data.

5.3.3 Physical-Layer Hybrid-ARQ Processing

The hybrid automatic repeat request (HARQ), the retransmission scheme adopted by LTE, is mainly a MAC sublayer procedure. However, the CRC check is done in the physical layer, and therefore part of the HARQ functionality is there, namely the transmission of ACK/NACK indications. Note that a HARQ retransmission is not simply a retransmission of a stored packet. If a retransmission is needed, then the packet must be reconstructed in the rate matching function.

TABLE 5.3 CHANNEL CODING SCHEMES AND CODING RATES FOR TRAFFIC CHANNELS

Traffic Channels	Coding Scheme	Coding Rate
UL-SCH	Turbo coding	1/3
DL-SCH		
PCH		
MCH		
BCH	Tail biting convolutional coding	1/3

TABLE 5.4 CHANNEL CODING SCHEMES AND CODING RATES FOR CONTROL INFORMATION

Control Information	Coding Scheme	Coding Rate
DCI	Tail biting convolutional coding	1/3
CFI	Block code	1/16
HI	Repetition code	1/3
UCI	Block code	Variable
	Tail biting convolutional coding	1/3

For the downlink data packets, the acknowledgments are sent on uplink PUCCH or PUSCH. For the uplink data, the same acknowledgments are sent on the downlink PHICH. See [2, 8] for details, which are rather complex.

5.3.4 Rate Matching and Channel Interleaving

The number of bits on a transport channel can vary with every transmission time interval—many applications generate variable bits rates. However, the allocated physical channel radio resources must be completely filled. The adjustment of the generated bit rate to the available bit rate is called *rate matching*.

Rate matching is depicted in Figure 5.9. The process is slightly different for turbo coded and convolutionally coded channels, but this top-level diagram applies to both cases.

The input to the rate matching is the output of the channel coding function. Three bitstreams are interleaved separately in a sub-block interleaver. The interleaver is a simple process whereby the bits are written into a matrix row by row, then the columns of the matrix are permutated according to a fixed pattern, and finally the bits are read again column by column. The matrix has always 32 columns, but the number of rows depends on the number of bits to be sent.

Next, the bit collection process simply combines the three bit streams that are the outputs of individual subblock interleavers, and the result is inserted into a circular buffer. Note that even at this stage of processing, the systematic bits are still in the beginning of the buffer, and the parity bits come in the end.

Bit selection and pruning is then the stage where the actual rate matching is done. Typically the number of bits from the channel coding process is larger than the allocated capacity, so some bits have to be punctured, or *pruned* (if there is more capacity than bits, then some bits can be duplicated). In principle, the systematic bits are more important than parity bits, and thus parity bits are what should be punctured. However, if excessive puncturing is applied to parity bits, the effective minimum Hamming distance

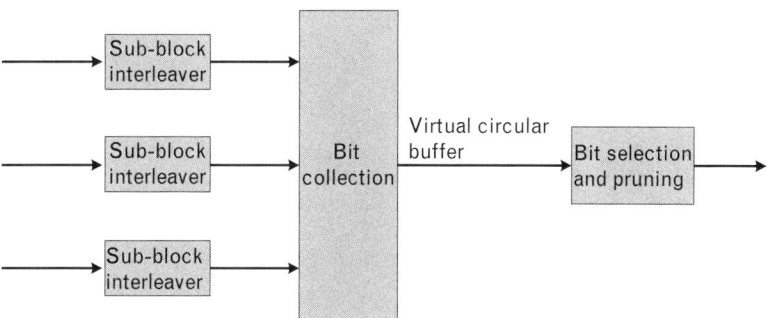

FIGURE 5.9
Rate matching.

of the punctured code can degrade at high code rates. Therefore, a small amount of systematic bit puncturing is adopted in LTE rate matching for turbo codes; this applies to about 6 percent of systematic bits.

Rate-matching function must also incorporate HARQ functionality. In case a negative ack was received, indicating that a data packet was received with errors, a retransmission must take place. The new transmission is not simply a retransmission of the old data packet, but the rate-matching function must make sure that each retransmission punctures different bits from the packet. This can easily be achieved with the help of the circular buffer. In case of a retransmission, the "starting point" for the bit selection and pruning function from the circular buffer is different from the original one, and this starting point changes with each successive retransmission. If the end of the buffer is reached, the reading of bits continues from the beginning of the buffer—thus the name *circular buffer*. From this we see that systematic bits do not get any preferential treatment in retransmitted packets, since they are no longer the first bits to be processed. The receiver will store all received packets, combine them, and try to decode the composite packet again. Therefore, the retransmission number (or redundancy version[RV], in the rate-matching process jargon) is an important parameter in the rate-matching function (see Figure 5.10). Note that this figure depicts basically the same process as in Figure 5.9, but in more detail. Rate matching is also explained in [3].

Note that in the beginning of this section, in CRC calculation, the transport block was divided into several code blocks in case there were more than 6,144 bits in a transport block. These code blocks went through channel coding and rate-matching processes separately, but after rate matching it is again time to combine the individual code blocks into a single codeword. This is done sequentially. It is also possible that the UE is configured to receive two transport blocks in a transmission time interval; in this case, the result is two codewords.

5.3.5 Scrambling

The purpose of scrambling is to convert the input bit string into a pseudo-random output bit string, thus removing as many long sequences of bits with similar values as possible. A long sequence of 1s or 0s might cause lots of problems at the receiver, especially the loss of timing synchronization and interference due to spectral spikes. See [1] for further information. This reference explains the generic scrambling used in the downlink. However, some physical channels employ their own channel-specific scrambling, while the uplink has its own scrambling procedures.

The scrambling generator is initialized at the start of each subframe; thus, scrambling does not have a memory over subframe boundaries.

5.3 PHYSICAL-LAYER PROCESSING

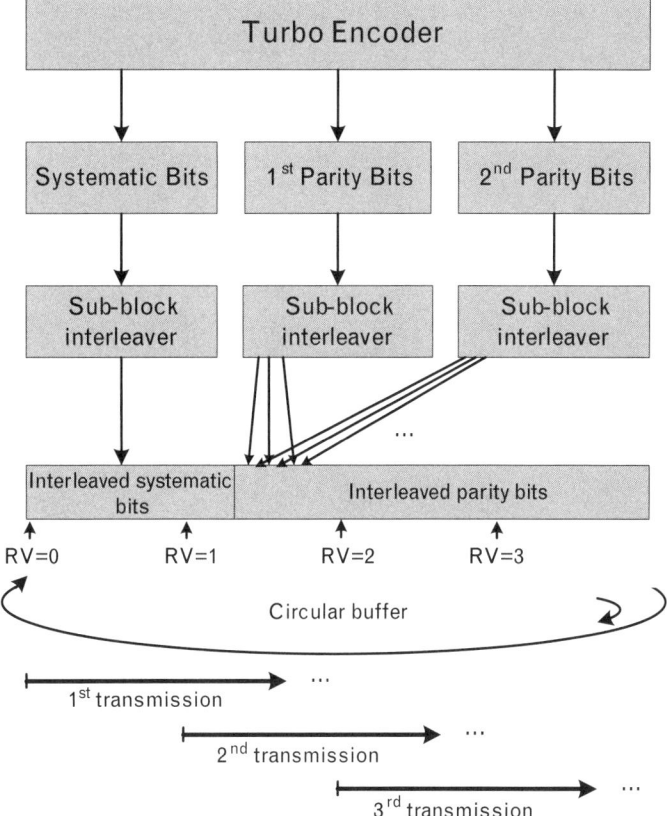

FIGURE 5.10
Rate matching and the circular buffer.

5.3.6 Modulation

It is easy to assume that in digital communication systems, we transmit bits, and each bit carries one piece of information in the form of one or zero. However, this is not the case. Modern wireless communication systems that employ digital modulation do not actually send digital bits; rather, they modulate an analog carrier signal in one way or another, and the items that are sent are called symbols. Each symbol can carry information from one or more bits, depending on the modulation scheme used.

A data-modulation scheme defines how the data bits are mixed with the carrier signal, which is always a sine wave. There are three basic ways to modulate a carrier signal in a digital sense: amplitude shift keying (ASK), frequency shift keying (FSK), and phase shift keying (PSK).

In ASK the amplitude of the carrier signal is modified (multiplied) by the digital signal. The modulated signal can be given as:

$$s(t) = f(t)\sin(2\pi f_c t + \phi) \quad (5.2)$$

where $s(t)$ is the modulated carrier signal and $f(t)$ the digital signal.

AIR INTERFACE: PHYSICAL LAYER

In FSK the frequency of the carrier signal is modified by the digital signal. If the digital signal has only two symbols, 0 or 1, this means that in the basic FSK scheme, the transmission switches between two frequencies to account for multilevel FSK. For mathematically minded people:

$$s(t) = f_1(t)\sin(2\pi f_{C1}t + \phi) + f_2(t)\sin(2\pi f_{C2}t + \phi) \qquad (5.3)$$

In PSK it is the phase of the carrier signal that is modified by the digital signal. Mathematically:

$$s(t) = \sin\left[2\pi f_C + \phi(t)\right] \qquad (5.4)$$

There are several variants in the PSK family. In binary phase shift keying (BPSK) modulation, each data bit is transformed into a separate data symbol. The mapping rule is 1 –> + 1 and 0 –> – 1. There are only two possible phase shifts in BPSK, 0 and π radians (see Figure 5.11).

The quaternary phase shift keying (QPSK) modulation has four phases: 0, 1/2 π, π, 3/2 π radians. Two data bits are transformed into one complex data symbol; for example, (00 –> + 1 + j), (01 –> – 1 + j), (11 –> – 1 – j), (10 –> + 1 – j). A symbol is any change (keying) of the carrier.

Generally, M-ary PSK has M phases, given as $2\pi m/M$; $m = 0, 1, \ldots, M - 1$.

Minimum shift keying (MSK) is a modification of QPSK. The GSM system uses Gaussian minimum shift keying (GMSK) modulation.

The number of times the signal parameter (amplitude, frequency, or phase) is changed per second is called the signaling rate or the symbol rate. It is measured in baud: 1 baud = 1 change per second. With binary modulations such as ASK, FSK, and BPSK, the signaling rate equals the bit rate. With QPSK and M-ary PSK, the bit rate exceeds the baud rate.

LTE uses four different modulation schemes: BPSK, QPSK, 16QAM (Figure 5.12), and 64QAM (Figure 5.13). The number of bit each symbol can carry in these schemes is given in Table 5.5.

FIGURE 5.11
BPSK and QPSK modulation.

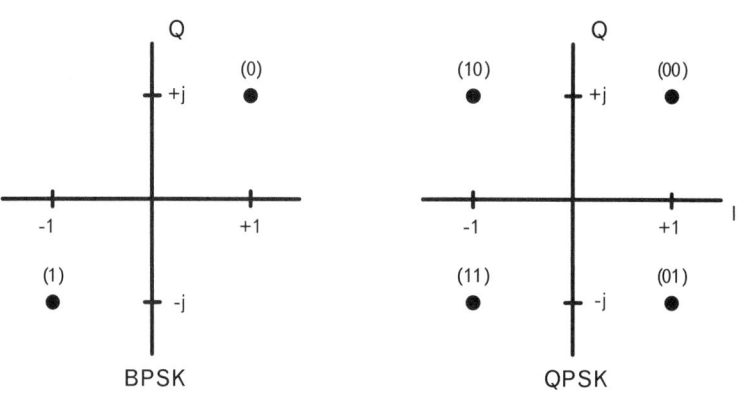

FIGURE 5.12
QPSK modulation.

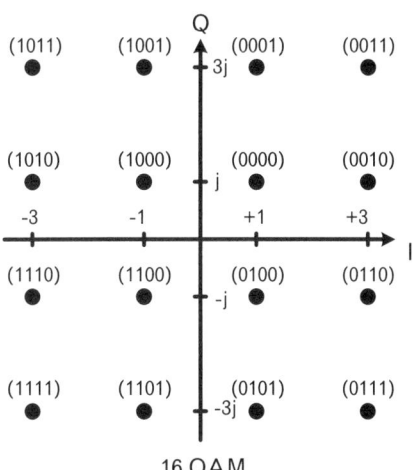

16 QAM

TABLE 5.5 LTE Modulation Schemes and Bits per Symbol

Modulation Scheme	Bits per Symbol
BPSK	1
QPSK	2
16QAM	4
64QAM	6

Note that the modulation schemes in LTE belong to two different modulation families. BPSK and QPSK are pure phase-shift keying modulation schemes, whereas 16QAM and 64QAM schemes combine features from both phase-shift and amplitude-shift keying schemes. BPSK and QPSK are more robust modulation schemes, and they can be used in situations where the amount of data transferred is small, but its error-free delivery is important. On the other hand, 16QAM and 64QAM can carry large amounts of information, but 64QAM especially requires very good channel conditions before it can be successfully employed.

BPSK is used only on some control channels: physical hybrid ARQ indicator channel (PHICH), and in some formats of physical uplink control channel (PUCCH).

QPSK is used on all other control channels where BPSK is not used. It provides a robust enough modulation for control channels that do not a require lots of transfer capacity. It can also be used on data channels if the radio conditions are difficult.

In addition, 16QAM and 64QAM are only used on data channels, and only when the radio conditions are good enough for their use. 64QAM is optional in the uplink. As seen from Figure 5.13, these schemes are error-prone in weak radio environment. For example, in 64QAM, it is hard to differentiate constellation points 000110 and 000111 at the receiver if the reception is not perfect.

The scheduling scheme used is decided by the scheduler at the eNodeB, and it can change dynamically even at every TTI. The value employed is signaled in a parameter called the modulation and coding scheme (MCS). For more information, see [1].

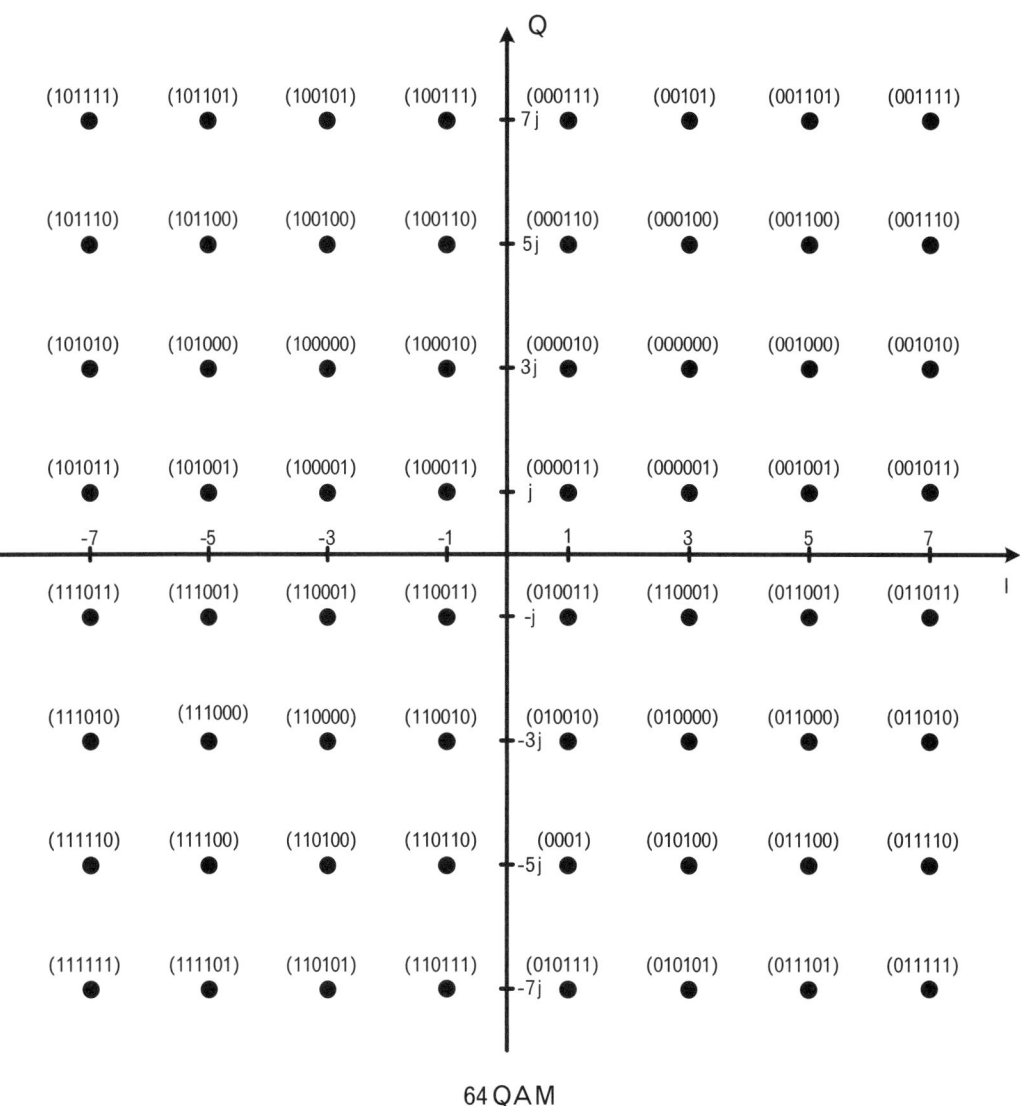

FIGURE 5.13 *64QAM modulation.*

5.3.7 Layer Mapping and Precoding

Layer mapping is a process of dividing the data so that it can be transmitted via several antennas in a multiantenna system. Note that a layer and an antenna are not synonymous. The number of layers can be less than or equal to the number of transmit antennas. Lets go through this phase by phase because understanding the relationship between layers and physical antennas is not clear even for many 3GPP specialists.

Layer is a somewhat abstract concept that defines how many simultaneous data streams there are to be transmittted. The number of layers can vary between 1 and 8. Layer mapping then defines how individual codewords (1 or 2) are mapped into layers. If there is only one layer, then this is easy: the single codeword is mapped onto the single layer. Note that it is not possible to have two codewords and a single layer. The number of codewords is always smaller than the number of layers, and the number of layers is always smaller than the number of antenna ports. A single codeword can be divided between (up to) four layers, and two codewords between (up to) eight layers. See Table 5.6 for clarification.

As a special case, if transmit diversity is employed instead of spatial multiplexing, then the number of codewords is one, and it can be mapped into two or four layers.

Precoding is the next process after layer mapping. Generally, in wireless systems, the capacity of the transmission link is much higher if the transmitter knows the channel state information and thus can adapt the transmission parameters accordingly (e.g., by selecting a suitable channel coding rate and

TABLE 5.6 MAPPING OF CODEWORDS TO LAYERS

LAYERS	CODEWORDS	MAPPING
1	1	The codeword is mapped to the single layer.
2	1	The codeword is divided evenly between the two layers.
2	2	Each codeword is mapped to a layer of its own.
3	1	The codeword is divided evenly between the layers.
3	2	The first codeword is mapped to the first layer, and the second codeword is divided evenly between layers 2 and 3.
4	1	The codeword is divided evenly between the layers
4	2	The first codeword is divided evenly between layers 1 and 2, and the second codeword is divided evenly between layers 3 and 4.
5	2	The first codeword is divided evenly between layers 1 and 2, and the second codeword is divided evenly between layers 3, 4, and 5.
6	2	The first codeword is divided evenly between layers 1, 2, and 3, and the second codeword is divided evenly between layers 4, 5, and 6.
7	2	The first codeword is divided evenly between layers 1, 2, and 3, and the second codeword is divided evenly between layers 4, 5, 6, and 7.
8	2	The first codeword is divided evenly between layers 1, 2, 3, and 4, and the second codeword is divided evenly between layers 5, 6, 7, and 8.

a modulation scheme). Usually this is accomplished by the receiver measuring the channel and then reporting the measurement results back to the transmitter.

With multiple antenna systems, this approach becomes increasingly difficult when the number of antennas increases. A 2x2 MIMO system has 4 individual component channels to measure and report, a 4x4 system has 16 channels, and an 8x8 MIMO system should report the channel state information from 64 channels. This is a large amount of data that should be reported at once, or it will quickly become outdated and thus useless for the transmitter.

A solution that has been adopted by LTE uses so called codebooks. A codebook is a collection of predefined matrices, each indicating a certain type of channel state (or actually the transmission parameters that are suitable for this channel state). The receiver selects one of them, one that is the closest to the measured channel state information, and only signals back the index of the selected matrix. Again, the system designer needs to make a compromise between how large the codebook is (and thus how accurately it signals the channel state information) and the number of bits needed to signal the matrix index in the codebook.

Once the precoder in the transmitter knows the precoding matrix index to be used, it simply takes the block of symbol vectors from the layer mapping as an input and multiplies it with the precoding matrix.

Precoding is a wide topic, and it is further discussed in the MIMO section in Section 11.2. Layer mapping and precoding are specified in [1].

5.3.8 Mapping to Assigned Resources and Antenna Ports

Antenna port is an unfortunate term in LTE. It easily gives an impression of a physical entity, a port, to a physical antenna. However, antenna port in LTE is a logical concept that identifies a channel with separate characteristics and planned usage. Each antenna port has been defined a separate reference signal (RS), such as a UE-specific RS for multilayer beamforming, or a UE-specific RS for single-layer beamforming. Reference [1] defines how layers are mapped into antenna ports. There are 27 antenna ports in the LTE-A Release 11 downlink. These are listed in Table 5.7.

Note that 3GPP specifications do not define how antenna ports are mapped to actual physical antennas. This is left for each base station manufacturer to decide.

5.4 Physical Channels in LTE-A

In this section we will discuss physical channels that are defined in LTE-A. Physical channels are used below the physical layer in the air interface, and

5.4 PHYSICAL CHANNELS IN LTE-A

TABLE 5.7 LTE Antenna Ports in the Downlink

Antenna Port Number	Downlink Reference Signal
0–3	Cell-specific RS
4	MBSFN-RS
5	UE-specific RS for single-layer beamforming
6	Positioning RS
7–8	UE-specific RS for dual-layer beamforming
9–14	UE-specific RS for multi-layer beamforming
15–22	Channel state information (CSI) RS
107–110	Demodulation RS associated with EPDCCH

they define the exact physical characteristics of radio channels. Different physical channels have different characteristics, and they are used to carry different types of data. A physical channel is a logical concept in the sense that there are no separate frequency resources for different physical channels; all of them are sent using the same frequency channel. However, each of them is allocated a different set of resource blocks.

Uplink and downlink have different set of physical channels, defined in [1, 3]. We will first introduce the downlink channels.

5.4.1 Downlink Physical Channels

Physical Broadcast Channel (PBCH)

- This channel carries the most important system information that the UE needs for accessing the cell. This information is defined in the Master Information Block (MIB).

- Other system information in System Information Blocks (SIB) is transmitted on the PDSCH.

- The data is modulated using QPSK.

- Channel coding is the standard LTE convolutional block coding.

- The scrambling sequence for PBCH is cell specific.

- The PBCH is mapped on the central 72 subcarriers of the frequency band.

- One PBCH transmission lasts 40 ms (i.e., the PBCH TTI is four radio frames).

Physical Control Format Indicator Channel (PCFICH)

- This channel informs the UE and the relay node about the number of OFDM symbols used for the PDCCHs (1, 2, 3, or 4).
- The channel coding is unique to PCFICH: 32,2 block coding that is defined in [3, Section 5.3.4].
- A PCFICH is transmitted on the first symbol of every subframe and carries a control format indicator (CFI) field.
- The PCFICH is modulated using QPSK.

Physical Downlink Control Channel (PDCCH)

- The physical downlink control channel carries scheduling assignments and other control information such as:
    - Downlink resource scheduling;
    - Uplink power control instructions;
    - Uplink resource grant;
    - Indication for paging or system information.
- The PDCCH carries the downlink control information (DCI) field, which can adopt several different formats: type 0, 1, 1A, 1B, 1C, 1D, 2, 2A, 2B, 2C, 2D, 3, 3A, and 4. This information can be UE specific, or it can apply to a group of UEs.
- Channel coding is the standard LTE convolutional block coding.
- Modulation is QPSK.

Enhanced Physical Downlink Control Channel (EPDCCH)

- This physical channel type was introduced only in Release 11, but it can co-exist on the same carrier as Rel-8/9/10 UEs.
- EPDCCH increases the downlink control channel capacity that might have become a bottleneck in LTE-A systems otherwise.
- EPDCCH carries information on the resource allocation of DL-SCH, hybrid ARQ information related to DL-SCH, and the uplink scheduling grant.
- Channel coding is the standard LTE convolutional block coding.
- The modulation scheme employed is QPSK.

5.4 PHYSICAL CHANNELS IN LTE-A

Physical Hybrid ARQ Indicator Channel (PHICH)

- Carries hybrid-ARQ ACK/NACKs in response to uplink data transmissions.
- Channel coding is simply repeating the ACK/NACK bit (0 or 1) three times (i.e., the resulting codeword is either 000 or 111).
- The modulation scheme employed is BPSK.

Physical Downlink Shared Channel (PDSCH)

- This is the main data-carrying channel; it carries the DL-SCH and the PCH.
- The channel coding is 1/3 rate turbo coding.
- Modulation can be QPSK, 16QAM, or 64QAM.
- Data transmitted on this channel goes through the whole processing chain that was explained earlier in this chapter.

Physical Multicast Channel (PMCH)

- This channel carries the MCH.
- No transmit diversity scheme is specified.
- Layer mapping and precoding will be done assuming a single antenna port (antenna port 4).
- This channel uses extended cyclic prefix.
- Physical layer processing for PMCH is very similar to PDSCH, including channel coding and modulation.

Relay Physical Downlink Control Channel (R-PDCCH)

- This channel informs the relay node about the resource allocation for DL-SCH.
- It carries hybrid ARQ information related to DL-SCH.
- It carries the uplink scheduling grant.
- The physical layer processing for R-PDCCH is very similar to PDCCH.

5.4.2 Uplink Physical Channels

The uplink has fewer physical channel types than the downlink. This is understandable because typically the intelligence in the system lies in the eNodeB, and it needs to send lots of different types of control information to UEs via the air interface. On the other hand, the uplink control information is mainly various indications and reports from UEs to the eNodeB.

Physical Uplink Control Channel (PUCCH)

- This is the only control channel type in the uplink. It can carry three different types of information:
  - Hybrid ARQ ACKs and NACKs in response to downlink transmissions;
  - Scheduling requests (SR);
  - Channel state information (CSI) reports (i.e., measurement reports).
- Like its downlink counterpart, PUCCH can adopt several different formats: 1, 1A, 1B, 2, 2A, 2B, and 3 (see Table 5.8).
- Channel coding depends on what kind of information the channel contains (i.e., its PUCCH format).
- The modulation scheme is BPSK or QPSK, depending on the PUCCH format.

Physical Uplink Shared Channel (PUSCH)

- This is the uplink data channel, carrying the UL-SCH.
- Channel coding is 1/3 rate turbo coding.
- The modulation scheme is QPSK, 16QAM, or optionally 64QAM.

Table 5.8 PUCCH Formats

PUCCH Format	Modulation Scheme	Number of Bits per Subframe
1	N/A	N/A
1a	BPSK	1
1b	QPSK	2
2	QPSK	20
2a	QPSK+BPSK	21
2b	QPSK+QPSK	22
3	QPSK	48

Physical Random Access Channel (PRACH)

- This channel carries the random access preamble (i.e., the request from the UE to initiate a connection).
- There are four different preamble formats (five in the TDD mode) and 64 different preambles in each cell.

5.5 Physical Layer Signals

The physical layer also has several different types of signals. The difference between a physical channel and a physical layer signal is that a physical channel carries information, whereas a signal is the information in itself. A physical layer signal is never mapped to a higher layer channel, so it does not carry information originating from higher layers. Signals are defined in [1].

5.5.1 Downlink Signals

LTE downlink can have two different types of signals:

- Reference signals;
- Synchronization signals.

5.5.1.1 Reference Signals

Reference signals are used for channel estimation. A reference signal transmits a known pattern, and by receiving this signal the UE can estimate what kind of imperfections the radio channel caused to the signal, and further it can exploit this information when receiving physical channels.

There are six different reference signals in the downlink:

- Cell-specific reference signals (CRS);
- MBSFN reference signals;
- UE-specific reference signals (DM-RS) associated with PDSCH;
- Demodulation reference signals (DM-RS) associated with EPDCCH;
- Positioning reference signals (PRS);
- CSI reference signals (CSI-RS).

There is one reference signal transmitted per downlink antenna port. The antenna ports and corresponding reference signals were given in Table 5.7.

The downlink cell-specific reference signal is transmitted in the first and the third-last OFDM symbol of each slot for antenna ports 0 and 1. If the reference signal is additionally transmitted from antenna ports 2 and 3, then the second OFDM symbol of each slot is used. The mapping of cell-specific reference signals to resource elements is depicted in Figure 5.14. Cell-specific reference signals are transmitted in all downlink subframes in a cell supporting PDSCH transmission.

MBSFN reference signals are transmitted in MBSFN subframes only when the PMCH is transmitted. MBSFN reference signals are transmitted on antenna port 4. They are spaced more "tightly" than other reference signals in the frequency domain (i.e., MBSFN reference signals are inserted in every other subcarrier), in the third, seventh, and eleventh OFDM symbols as shown in Figure 5.15. MBSFN reference signals are defined for the extended cyclic prefix only.

UE-specific reference signals can be transmitted on antenna port(s) 5, 7, 8, 9, 10, 11, 12, 13, and 14. The number of antenna ports depends on the layers used for the transmission of the PDSCH. The mapping of UE-specific reference signals is rather complex. The example in Figure 5.16 is for antenna port 5. Here the reference signals are transmitted on the fourth and seventh OFDM symbol in even-numbered slots, and on the third and sixth OFDM symbol in odd-numbered slots. Other antenna ports and frames with extended cyclic prefixes use different mapping as explained in [1].

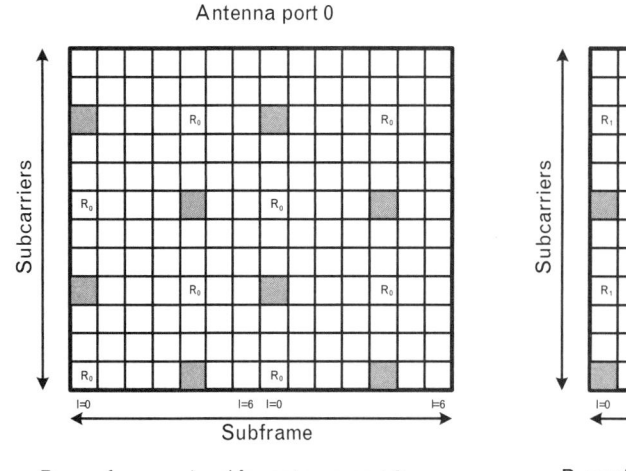

FIGURE 5.14 *Cell-specific reference signal mapping.*

FIGURE 5.15
MBSFN reference signal mapping.

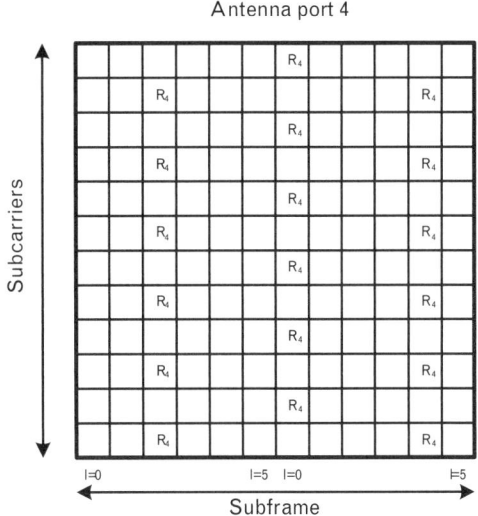

R_4 = reference signal for antenna port 4

Demodulation reference signals associated with EPDCCH are a new reference signal type, only introduced in Release 11. They can be transmitted from antenna ports 107, 108, 109, and 110; the exact number depends on how many layers are used. The mapping of reference signals to resource elements for antenna ports 107–110 is similar to mapping UE-specific reference signals for antenna ports 7–10.

Positioning reference signals (PRSs) are transmitted on antenna port 6 as in Figure 5.17. They are used to improve the accuracy of the observed

FIGURE 5.16
UE-specific reference signal mapping; antenna port 5 and normal cyclic prefix.

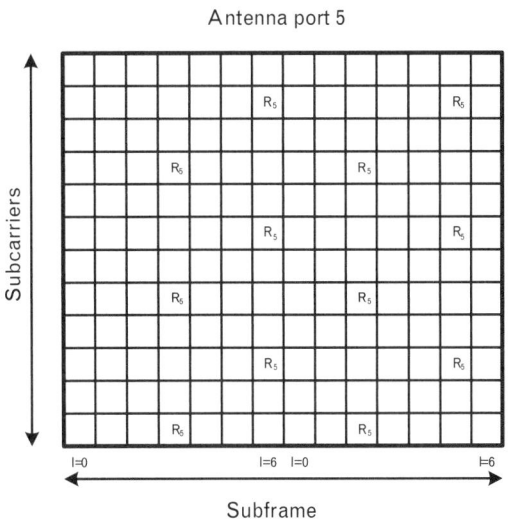

R_5 = reference signal for antenna port 5

FIGURE 5.17
Positioning reference signal mapping.

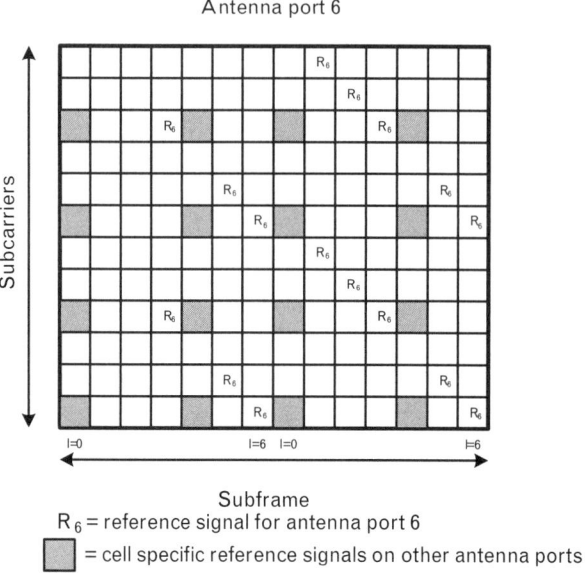

time difference of arrival (OTDOA) positioning method. In this positioning method, the UE position is calculated based on the measured time difference of the arrival of reference signals from the serving cell and at least two neighboring cells. This sounds simple, but the problem is that the UE may not hear the standard reference signals from neighboring cells, especially if the UE is in the middle of the serving cell. Therefore positioning reference signals were introduced in Release 9. The purpose of PRSs is to improve the "hearability" of neighboring cells. PRS subframes do not overlap with PDCCH or cell-specific reference signals, and their transmission is not scheduled on the same subframes as PDSCH. PRSs are also transmitted on several successive subframes to improve their detection. Moreover, with the help of cell-specific frequency shifts, it is possible to define six different nonoverlapping PRS patterns for neighboring cells. This is also a very useful feature in increasing PRS hearability. If, despite this feature, there are still overlapping PRSs in neighboring cells, the network can order some cells to halt their PRS transmissions for a number of subframes. During these subframes, the hearability of the remaining (now nonoverlapping) PRS will improve. The positioning reference signals are not mapped to resource elements that are already allocated to the PBCH, primary synchronization signal (PSS), or secondary synchronization signal (SSS) regardless of their antenna port.

Channel state information (CSI) reference signals are used for obtaining CSI estimation from up to eight antenna ports. CSI reference signals are transmitted on one, two, four, or eight antenna ports.

5.5.1.2 Synchronization Signals.

Synchronization signals are the first signals the UE needs to receive if it wants to camp into the network. The UE needs to receive PBCH in order to get vital information on the system. However, before it can do that, it has to synchronize itself with the eNodeB.

There are two types of synchronization signals: the primary synchronization signal and the secondary synchronization signal. The PSS provides slot timing detection, and the SSS provides the radio frame timing detection. The synchronization signals are sent periodically, twice in a 10-ms radio frame. The PSS is sent in the last OFDM symbol of the first and eleventh slots of each radio frame. The SSS is transmitted in the preceding OFDM symbol, in both PSS occurrences.

The synchronization procedure itself is explained in Section 9.2 and that section also includes a more thorough handling of synchronization signals.

5.5.2 Uplink Signals

In the uplink direction there are only two reference signal types:

- Demodulation reference signal, associated with the transmission of PUSCH or PUCCH;
- Sounding reference signal, not associated with the transmission of PUSCH or PUCCH.

The demodulation reference signal is associated with transmission of PUSCH data or PUCCH control information. This reference signal is provided, as its name suggests, for channel estimation so that the eNodeB can demodulate the received data or control channel. There is one demodulation reference signal in every uplink PUSCH or PUCCH slot.

In case of PUSCH (and normal cyclic prefix), the demodulation reference signal is added to the third symbol of each slot. In case of PUCCH, the situation is more complex because PUCCH itself does not have a standard single format but several PUCCH formats have been defined. Table 5.9 shows the demodulation reference signal location for different PUCCH formats.

Sounding reference signals (SRS) are not associated with any particular channel. Instead they are used by the network to get channel quality information, which is then used as a parameter in frequency-selective scheduling. The SRS, if used, is added in the last SC-FDMA symbol in a subframe. A SRS transmission is triggered by a request from the eNodeB. The request can indicate a single SRS transmission or periodic SRS transmissions.

TABLE 5.9 DEMODULATION REFERENCE SIGNAL LOCATION FOR VARIOUS PUCCH FORMATS

PUCCH Format	SC-FDMA Symbol Number(s)	
	Normal Cyclic Prefix	Extended Cyclic Prefix
1, 1a, 1b	2, 3, 4	2, 3
2, 3	1, 5	3
2a, 2b	1, 5	N/A

5.6 Summary

This chapter discussed the LTE air interface physical layer. We omitted OFDM-specific issues since those were already discussed in Chapter 4. Instead, this chapter concentrated on physical layer processing, which is for the most part surprisingly independent of the air interface access technology used.

The chapter started with a general overview of the physical layer, including the radio frame structure for both FDD and TDD modes. However, the main content of the chapter is in Section 5.3, Physical Layer Processing. A diagram is presented, showing the main processing stages for downlink data packets received from higher layers, and then these stages are discussed one by one in separate subsections. Physical channels and physical layer signals were also presented in Sections 5.4 and 5.5, respectively.

The presentation in this chapter is mostly FDD-mode specific. However, the TDD mode in LTE is purposely kept very similar to the FDD mode whenever this has been possible. As proof of this, there are no separate sets of FDD and TDD mode physical layer specifications in LTE, whereas in UMTS we had 25.21x series specifications for the FDD mode, and 25.22x series for the TDD mode. In LTE, the 36.2xx series covers both modes.

It must again be stressed that the air interface physical layer is a very complex entity in LTE. This chapter discusses only the principles of the physical layer. The details and various optional functionality can then be studied from the specifications itself once the principles are understood. Also, because of Section 5.3, the discussion in this chapter is downlink-centric. The physical layer in the uplink is different from the downlink, but the same principles apply to both. If the downlink is well understood, then there should be no problems with the uplink either.

References

[1] 3GPP TS 36.211, v 11.4.0, Evolved Universal Terrestrial Radio Access (E-UTRA); Physical Channels and Modulation, 09/2013.

5.6 SUMMARY

[2] 3GPP TS 36.300, v 11.7.0, Evolved Universal Terrestrial Radio Access (E-UTRA) and Evolved Universal Terrestrial Radio Access Network (E-UTRAN); Overall Description; Stage 2; 09/2013.

[3] 3GPP TS 36.212, v 11.3.0, Evolved Universal Terrestrial Radio Access (E-UTRA); Multiplexing and Channel Coding, 06/2013.

[4] Berrou, C., A. Glavieux, and P. Thitimajshima, "Near Shannon Limit Error Correcting Coding and Decoding: Turbo-Codes (1)," *Proc. IEE Int. Conf. on Communications*, Geneva, Switzerland, May 1993, pp. 1064–1070.

[5] Shannon, C. E., "A Mathematical Theory of Communication," *Bell Systems Technical Journal*, Vol. 27, 1948, pp. 379–423, 623–656.

[6] Goldsmith, A., *Wireless Communications*, Cambridge: Cambridge University Press, 2005, pp. 362–364.

[7] Viterbi, A. J., *CDMA: Principles of Spread Spectrum Communication*, Reading, MA: Addison Wesley Longman, 1995.

[8] 3GPP TS 36.213, v 11.4.0, Evolved Universal Terrestrial Radio Access (E-UTRA); Physical Layer Procedures; 09/2013.

CHAPTER 6

Air Interface: Protocol Stack

6.1 Introduction

This chapter continues from where the previous chapter ended. The air interface protocol stack uses the services provided by the physical layer.

The protocol stack architecture is typically considered to consist of two different parts: control and user planes. The control plane carries control data (i.e., signaling data that is needed for the management of the UE connection and indeed for many tasks that take place already before a connection is established). The user plane carries user data that originates from various applications, including voice calls. Whereas both planes have largely the same protocol tasks, their functions are very different. Particularly in LTE, the data rates can be very high, and therefore user plane functions have to be designed (and, especially, implemented) in a way that enables high data throughput.

Sometimes the air interface protocols are also divided into access stratum (AS) and nonaccess stratum (NAS) protocols (see Figure 6.1). LTE has inherited this concept from UMTS. In principle, AS protocols are those that handle the radio access between the UE and the eNodeB. These protocols terminate in the RAN. The NAS includes core network (CN) protocols between the UE and the CN itself. These protocols are not terminated in the RAN, but in the CN; the eNodeB is transparent to the NAS. In principle, NAS protocols are independent of the radio access technology below them—though in case of LTE this principle remains unexploited—the LTE access technology is only deployed with the LTE core network [1].

Since this chapter is about the radio interface, we will only concentrate on AS protocols here. NAS and NAS procotols (ESM and EMM) are briefly discussed in the core network chapter.

Figures 6.2 and 6.3 present the protocol layers in the air interface [2]. They are introduced one by one in following sections, in each case discussing both the user plane and control plane functionality. Note that radio resource control (RRC) is a pure control plane entity.

90 AIR INTERFACE: PROTOCOL STACK

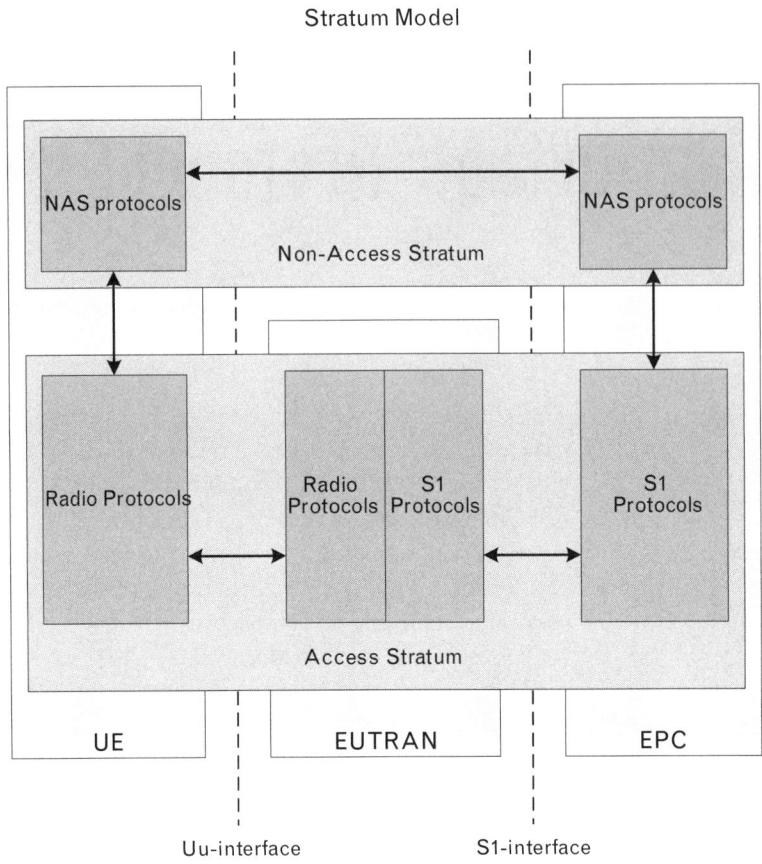

FIGURE 6.1
Stratum model.

6.2 Medium Access Control (MAC)

6.2.1 MAC—General

MAC is the protocol task closest to the physical layer [3]. It works closely with the physical layer in many processes such as HARQ and random access. Indeed, the model of the MAC layer presented in 3GPP specifications and in this book is just a model. An equipment vendor is free to implement the physical layer and the MAC layer in a way it considers to be the most appropriate and efficient solution. The interface between the PHY and the MAC layers is not an open interface but a vendor-specific one. The functionality of these entities can be moved around, as long as the air interface conforms with the standard and remains an open interface. Because of its close relationship with the physical layer, many of MAC's processes are tightly time constrained. MAC is configured by the RRC layer.

The MAC protocol is not a symmetrical protocol; the entities in the UE and in the eNodeB are different. This is due to the fact that in many ways, the UE acts as a slave and the eNodeB as a master in this relationship.

6.2 MEDIUM ACCESS CONTROL (MAC)

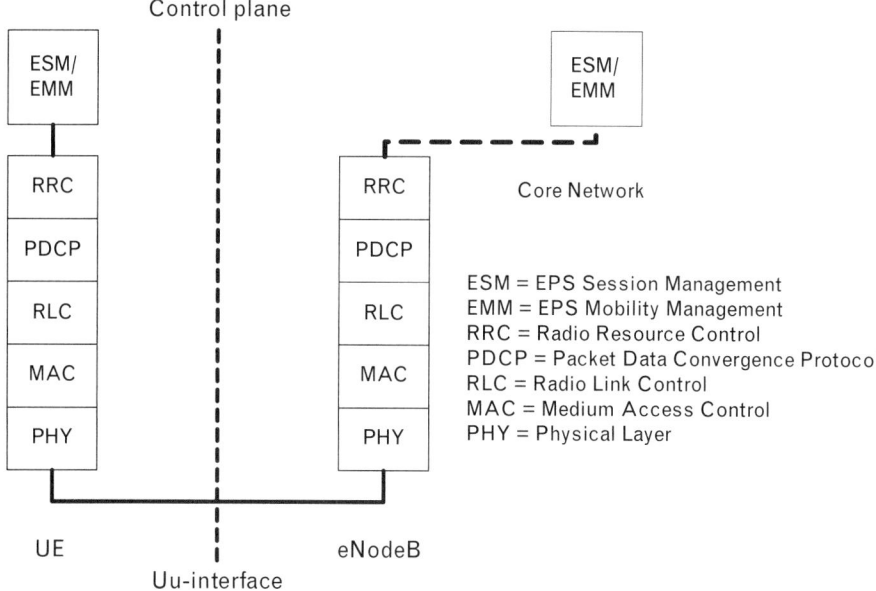

Figure 6.2 *Control plane protocol stack.*

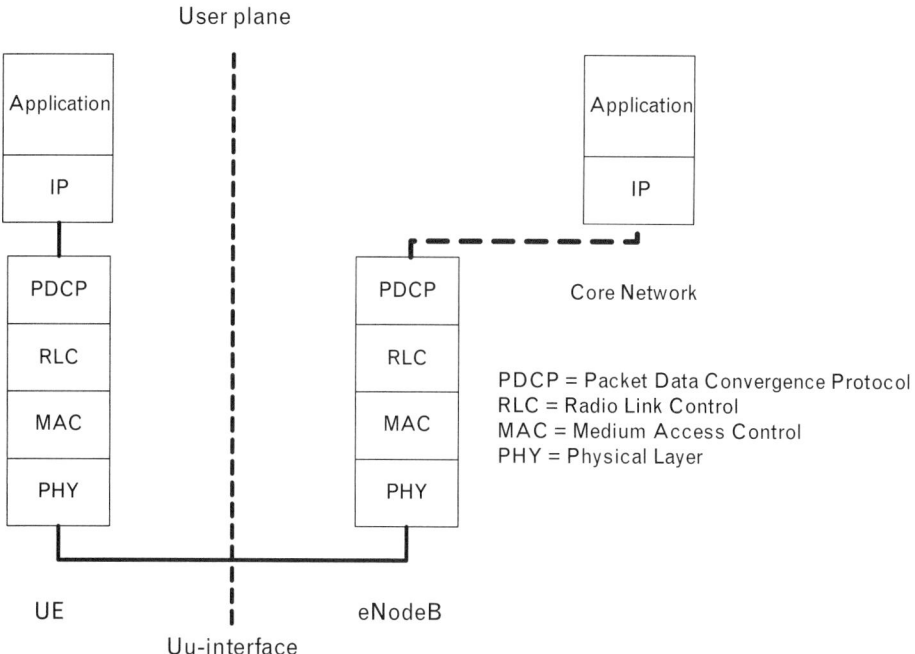

Figure 6.3 *User plane protocol stack.*

The UE provides measurement and scheduling information to the eNodeB, and, based on the information available, the eNodeB manages the UE. The MAC functionality in the UE is closely standardized, but the MAC layer specifications for the eNodeB have been left much more open. This is because the MAC scheduler is an especially important factor for the whole network performance. It was not seen as appropriate to restrict the implementation of the eNodeB by standards that were too detailed. Instead the equipment vendor and/or the network operator can fine-tune the eNodeB MAC layer more freely. Similar approach was not chosen for UEs because of the large number of different UE vendors and devices. That is, a network operator can trust itself to implement the eNodeB MAC properly, but it cannot trust that hundreds of UE manufacturers will do the same.

Note that relay nodes (RN) include the functionality of both the UE and the eNodeB MAC, since it needs the former to communicate with the eNodeB and the latter to communicate with the UE [2, 4].

Readers familiar with UMTS may recall that UMTS-MAC had several functional entities, such as MAC-b and MAC-c. LTE does not follow this convention anymore. Instead, Figure 6.4 shows a possible MAC structure at the UE side.

The following functions are supported by the MAC protocol layer:

- Mapping between logical channels and transport channels;
- Multiplexing of MAC SDUs;
- Demultiplexing of MAC SDUs;
- Scheduling information reporting;
- Error correction through HARQ;
- Priority handling between UEs;
- Priority handling between logical channels of one UE;
- Logical channel prioritization;
- Padding;
- Transport format selection.

Note that this list includes MAC functionality both in the UE and in the eNodeB. Table 6.1 shows how these functions are divided between the UE and the eNodeB, and between downlink and uplink.

This list does not include all tasks done by a MAC layer. A given UE (or eNodeB) function typically involves several protocol tasks, and sometimes it is difficult to define which layer "owns" this functionality.

We will briefly explain the meaning of the functions from Table 6.1 in following sections.

6.2 MEDIUM ACCESS CONTROL (MAC)

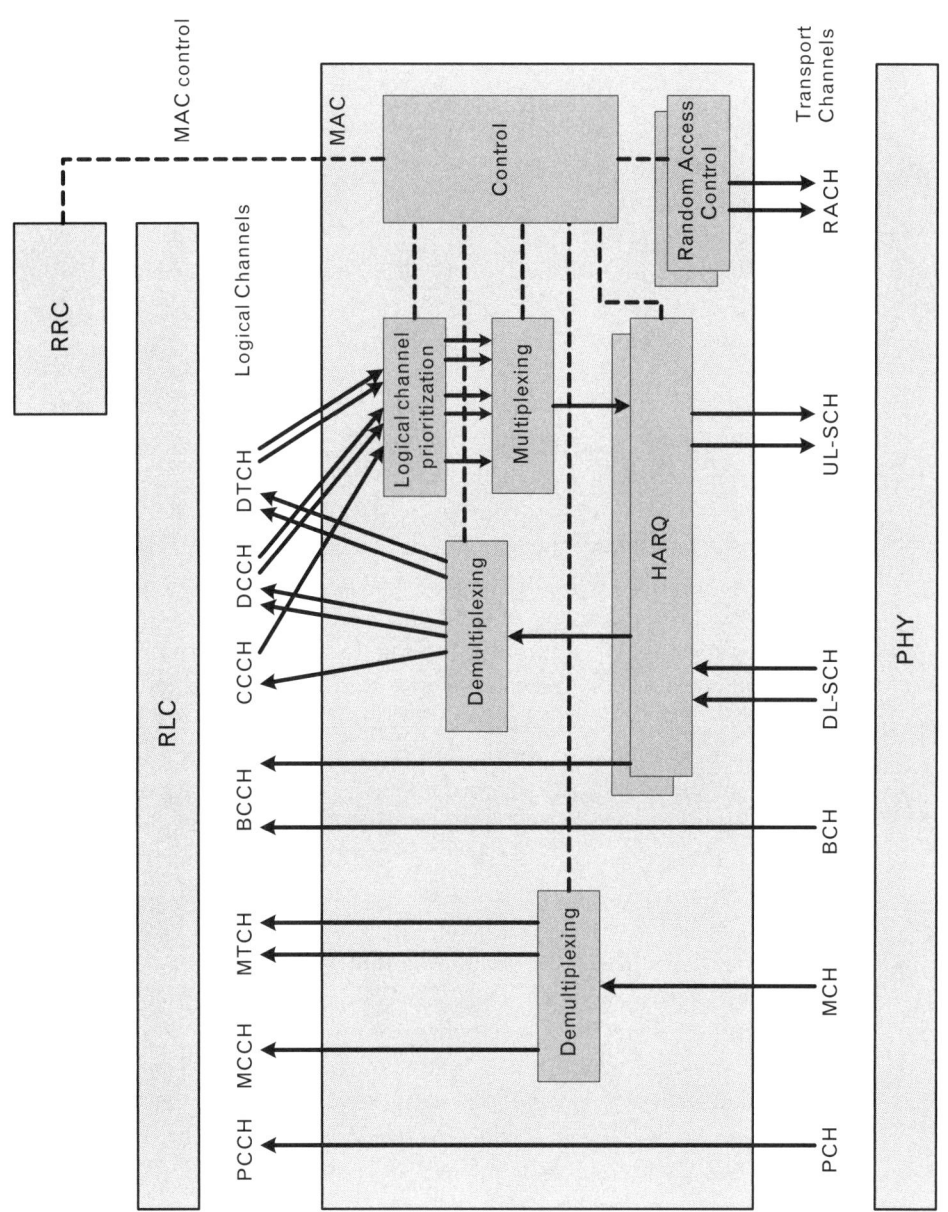

FIGURE 6.4 *MAC structure, UE side.*

TABLE 6.1   MAC Functionality in the UE/eNodeB

MAC Function	UE	eNodeB	Downlink	Uplink
Mapping between logical channels and transport channels	X		X	X
		X	X	X
Multiplexing	X			X
		X	X	
Demultiplexing	X		X	
		X		X
Error correction through HARQ	X		X	X
		X	X	X
Transport format selection		X	X	X
Priority handling between UEs		X	X	X
Priority handling between logical channels of one UE		X	X	X
Logical channel prioritization	X			X
Scheduling information reporting	X			X
Padding	X			X
		X	X	

Multiplexing means combining data from several different (or indeed from the same) logical channels onto transport blocks (TB) to be delivered to the physical layer on transport channels. Demultiplexing is its mirror process at the other end of the transmission link. The channels that can be multiplexed together are shown in Figure 6.4. The actual multiplexing configuration (i.e., how much data and from which channels are multiplexed together and transmitted in each TTI) is controlled by the RRC layer.

Logical channel prioritization takes place in the uplink direction. The RRC layer assigns priority parameters for each logical channel, and the scheduling of uplink data is based on these parameters; the channels with higher priority are served first. This is a two-stage process: in the first round each channel is allocated capacity based on its prioritized bit rate (PBR) value, in the order of the channel priority parameter. If, after all channels with buffered data have been allocated their PBR capacity, there is still unused capacity left, then the UE will allocate that capacity in the channel priority order until all capacity has been used. This two-stage approach tries to prevent a situation where a lower-capacity channel never gets its data through if there is data buffered in higher-priority channels. See an example in Figure 6.5. The data is added to the MAC PDU in the order shown; first the PBR data in blocks numbered 1–4, and then additional non-PBR data in blocks numbered 5 and 6.

Transport format selection is a process that happens in the eNodeB only, for both the downlink and the uplink. After selecting a suitable transport format, the eNodeB then sends the information to the UE, to be used in the uplink transmissions. Transport format defines with what kind of

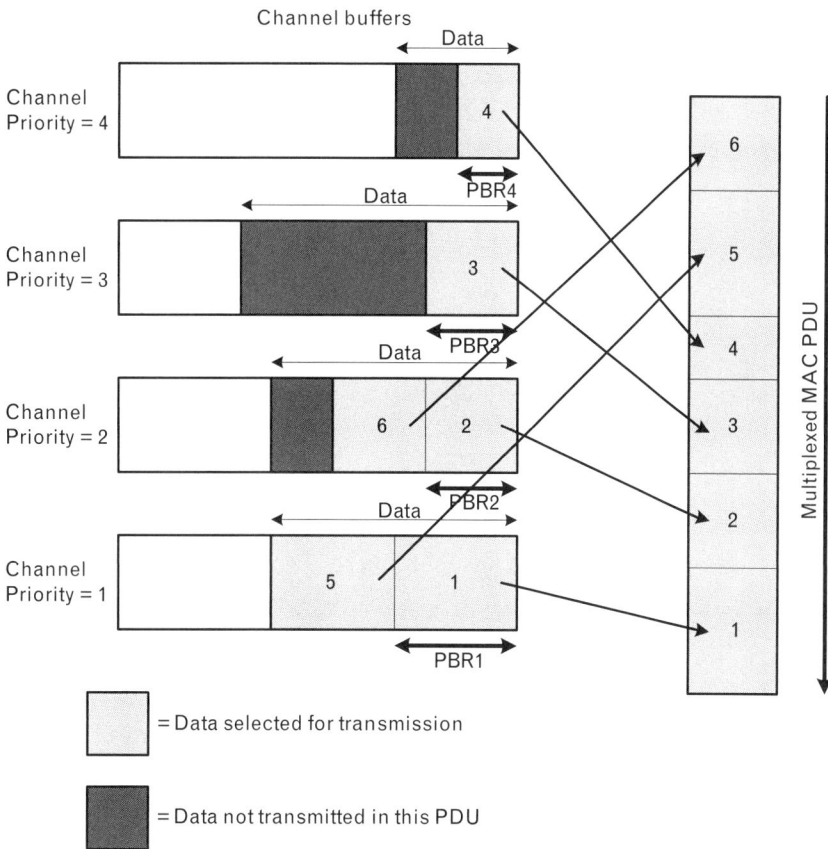

FIGURE 6.5
Logical channel prioritization.

characteristics the data is sent over the air interface. Transport format indicates parameters such as channel coding, modulation, and interleaving. In fact, the term *transport format* is rarely used in LTE specifications (whereas this term is everywhere in UMTS standards). An interested reader looking for further information is advised to look instead for terms *downlink control information* (DCI) and especially *modulation and coding scheme* (MCS) within the DCI. DCI comes in many different formats, which are suitable for different situations. Uplink grants (and MCS) and sent using DCI formats 0 or 4. Format 0 is for normal channels; format 4 for MIMO channels. For the downlink, there are several DCI formats [5]. DCI is carried on the physical downlink control channel (PDCCH). In the downlink the new transport format received via DCI applies to the next transport block, but for uplink the DCI 0/4 applies only to subframe n+4 where n is the current subframe. This delay is necessary so that the UE has some time to process the outgoing data appropriately. Note that the new DCI information will not necessarily be sent in every subframe. DCI allocation can also be semipersistent. Semipersistent scheduling can take place when the same amount of data is sent from subframe to subframe. In this case, the eNodeB can configure the

UE to use the same transport format until advised otherwise. Semipersistent scheduling is expected to be used with the VoIP service especially.

Scheduling information reporting is an important task for the UE. It has to provide the eNodeB with information that can be used in eNodeB's MAC scheduler. This information includes the amount of data in UE's buffers and other information added to MAC PDU headers.

If there is not enough data to use all the allocated resources, then padding will be used to fill the PDU with dummy bits.

6.2.2 Transport Channels

Transport channels define how and with which type of characteristics the data is transferred by the physical layer. These channels are used in the interface between the MAC and the physical layer. All transport channel types are unidirectional; they exist either in the downlink or in the uplink but not in both.

The mapping of transport channels to physical channels is pretty straightforward (see Figures 6.6 and 6.7). These figures depict the situation from the UE point of view (i.e., they show the UE uplink and the UE downlink channel mapping) [5].

As seen, some of the physical channels do not map to any transport channels. Instead, they are transport control information such as uplink control information (UCI), downlink control information (DCI), hybrid ARQ indicator (HI), and control format indicator (CFI).

6.2.2.1 Downlink Transport Channels

Broadcast Channel (BCH)
The BCH carries the most important system information, the master information block (MIB), which informs the UE about the configuration of the cell. This information is needed for the UE so that it can access the cell,

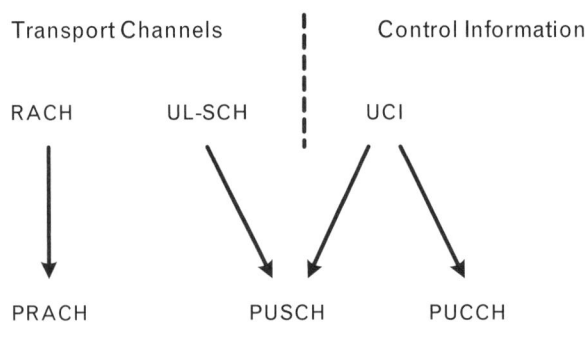

FIGURE 6.6
Transport channel mapping—uplink.

6.2 MEDIUM ACCESS CONTROL (MAC)

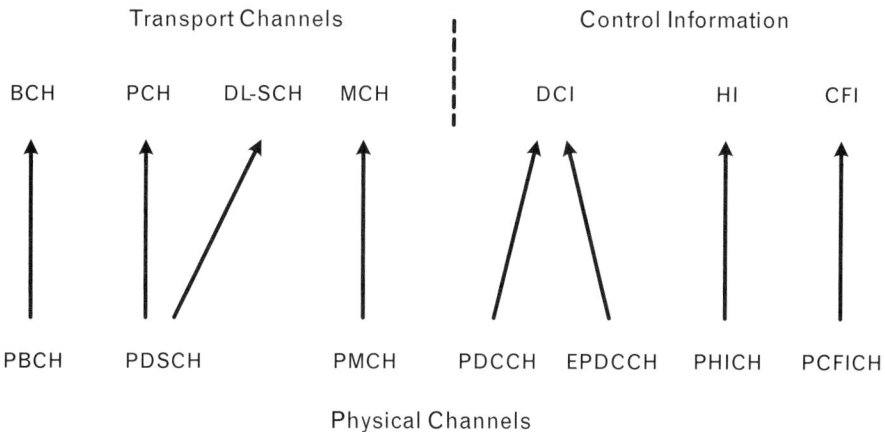

FIGURE 6.7   *Transport channel mapping—downlink.*

read more system information via other channels, and set up a connection. Without a successful reception of the MIB, the UE cannot use any services of the cell. The transport format used in this channel is static so the UE does not need to decode anything else in order to receive the MIB. There is also a requirement that this channel has to be broadcast over the whole cell area.

The BCH channel maps to PBCH in the physical layer.

Downlink Shared Channel (DL-SCH)
This transport channel is the main data transfer channel for the downlink. DL-SCH can carry several different logical channel types, including both data and control channels. Notably, the non-MIB parts of the system information, the system information blocks (SIB), are transmitted on DL-SCH. This channel can use the whole toolbox of MAC procedures and functions in order to provide efficient data transfer services, including HARQ, adaptive modulation and coding (AMC), discontinuous reception (DRX), and multiple input multiple output (MIMO) antenna techniques.

This channel maps to PDSCH in the physical layer.

Paging Channel (PCH)
Like the BCH, the PCH also needs to be transmitted over the whole cell area. Paging channel carries paging information (i.e., notifications of incoming user data). In order to receive this user data, the UE has to set up a connection using the random access procedure. The UE should monitor the PCH regularly so that it does not miss any incoming connection requests. However, continuous monitoring is not necessary because PCH supports the DRX scheme whereby the PCH channel for this particular UE is sent only on certain times. This is an important feature that saves power and in-

creases the UE standby time because the UE can go to sleep mode outside the scheduled PCH transmission periods.

In addition to connection requests, the PCH also carries indications that the system information has been modified and thus the UE should re-read it from the DL-SCH, or that a new public warning system (PWS) message is available via SIB on the DL-SCH. PWS carries, for example, warnings on natural disasters. These warnings are sent on SIB 10/11 in case of the Earthquake and Tsunami Warning System (ETWS, the Japanese PWS system), or on SIB12 in case of the Commercial Mobile Alert System (CMAS, the North American System), the Korean Public Alert System (KPAS, the Korean system), or the EU-Alert (the European Union system).

This channel maps to PDSCH in the physical layer (i.e., there is no dedicated physical paging channel).

Multicast Channel (MCH)
This transport channel carries MBMS information, either control or user data, that requires MBSFN combining (i.e., the data is broadcast over multiple cells). Note that the other type of MBMS data ("the single-cell MBMS") can be transmitted over DL_SCH.

This channel maps to PMCH in the physical layer.

Note that there is no dedicated MAC layer downlink control channel. MAC control information is received via higher layer (RRC) signaling or via MAC PDU headers.

6.2.2.2 Uplink Transport Channels

Uplink Shared Channel (UL-SCH)
This transport channel is the main channel for uplink data transfer. In fact, all uplink logical channels use UL-SCH as their transport channel. Like its downlink counterpart, UL-SCH can employ HARQ, AMC, and MIMO.

This channel maps to PUSCH in the physical layer.

Random Access Channel (RACH)
This channel is used for random access. It has to be used when the UE still does not have uplink timing synchronization or allocated uplink resources. RACH can carry only a very limited amount of data, and because of the type of the channel there is always a risk of collisions with RACH transmissions from other UEs.

This channel maps to PRACH in the physical layer. Transport channels are also discussed in [2].

6.3 Radio Link Control (RLC)

6.3.1 RLC—General

The RLC task is simpler than MAC, at least when it comes to functionality. In general, the RLC layer is in charge of the actual data packet, containing either control or user data transmission over the air interface. It makes sure that the data to be sent over the radio interface is packed into suitably sized packets. The RLC task maintains a retransmission buffer, performs HARQ reordering, and routes the incoming data packets to the right destination task. Depending on the type of data, there are three alternative services it can provide via its functional entities:

- Transparent mode (TM);
- Unacknowledged mode (UM);
- Acknowledged mode (AM).

One RLC task contains several different functional entities. For data using the transparent mode service or the unacknowledged mode service, there is one transmitting and one receiving entity for each service. For data using the acknowledged mode service, there is only one combined entity handling both the transmission and reception.

The transparent mode is used for the BCCH, PCCH, and CCCH (both uplink and downlink) channels. Transparent mode means that very little processing is done to the data in the RLC. The transmitting TM entity contains only a transmission buffer, and the receiving task does not even have that. Note that no RLC header is added to data units in the transparent mode (see Figure 6.8).

The transparent mode is used only to transport higher layer control data that does not require additional protection from the RLC layer. This data includes broadcast and paging messages, and control messages sent using the CCCH. The transparent mode is not used for user data.

The UM mode is used for the DTCH (both uplink and downlink), MCCH, and MTCH channels. This entity either segments or concatenates upper layer packets so that they are of suitable size for transmission. The RLC also adds a header to the PDU; this header includes the sequence number of the packet to be used for HARQ reordering at the receiver (see Figure 6.9).

The UM mode is used for delay-sensitive real-time applications. For such services, the AM mode cannot be used because of the delays that the additional data processing, especially packet retransmissions, would cause. Also, point-to-multipoint services such as MBMS will use the UM mode.

100 AIR INTERFACE: PROTOCOL STACK

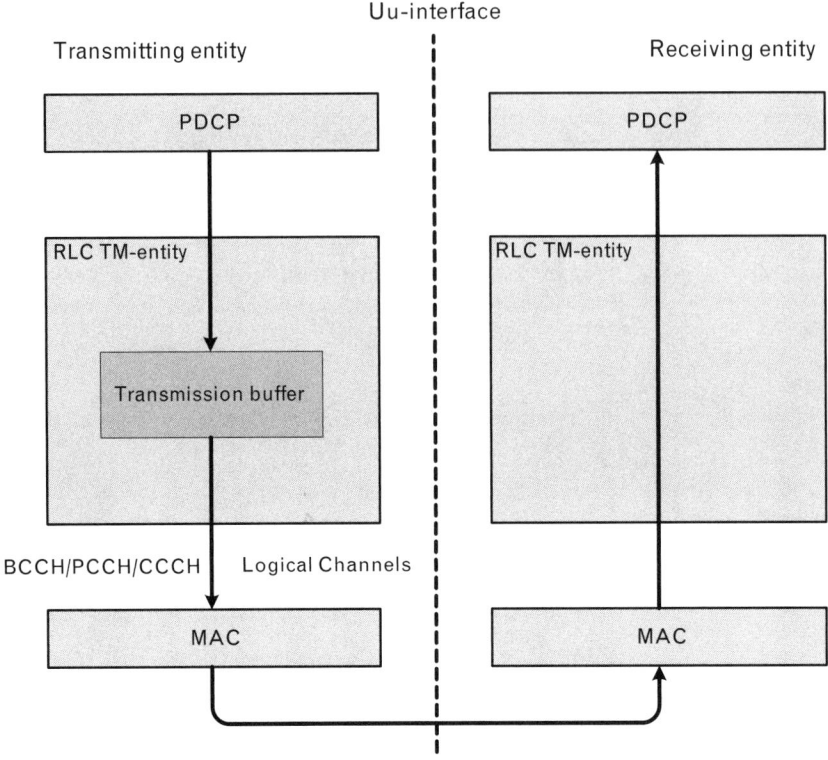

FIGURE 6.8
Transparent mode entity in RLC.

The AM mode can be used for the DCCH and DTCH channels. The AM entity is bidirectional; the same entity can handle both the transmission and the reception of data. As seen from Figure 6.10, the major difference between UM and AM modes is the added retransmission functionality in the AM mode. The AM mode aims to provide error-free data transmission service by means of retransmissions in case of incorrectly received packets. However, a retransmission (or several retransmissions) of the same data packet increases the delay, and thus the AM mode is more suitable for delay-tolerant services such as file downloads, email, and web browsing. Also streaming services can use the AM mode if some amount of delay is accepted. Streaming services should employ a large buffer at the receiver to mitigate the effects of jitter (which will inevitably occur as a result of retransmissions).

Once the AM entity sends a data packet, it is also stored into the retransmission buffer. If the transmitting entity receives a positive acknowledgment, the packet is disposed of the buffer. In case a retransmission is needed, the packet can be retransmitted as such or, if the MAC layer indicates that the packet size will be smaller for the retransmission, the buffered packed can be resegmented. Typically a retransmitted packet will have a smaller data content than the original one, not a larger one. This is because

6.3 RADIO LINK CONTROL (RLC) 101

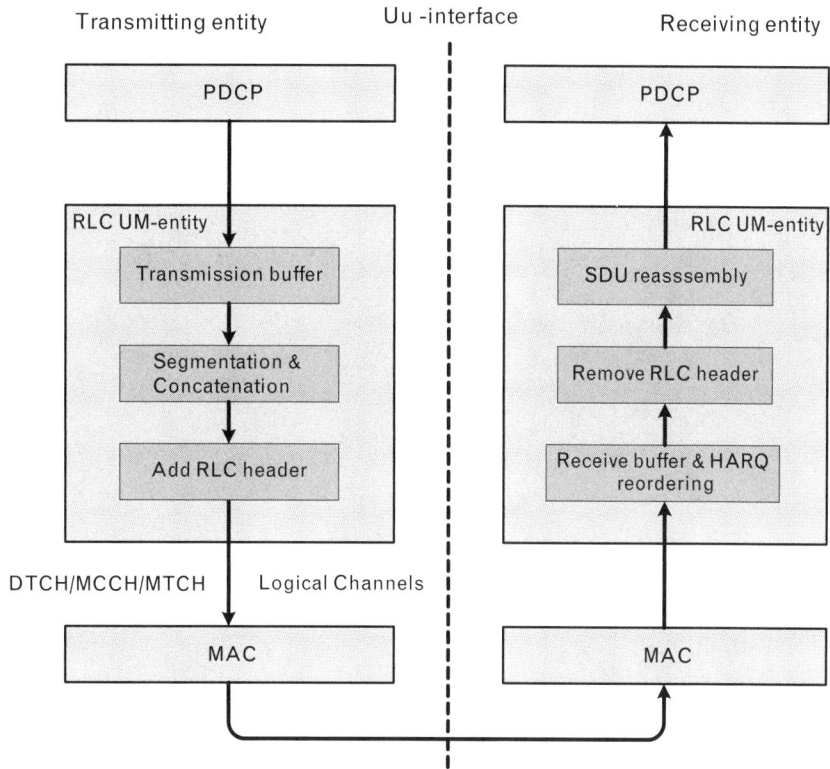

FIGURE 6.9 Unacknowledged mode entity in RLC.

if a transmission failed, it is possible that the MAC will employ more robust transmission parameters such as the modulation scheme, and therefore less data can be sent within a similar packet size.

An RLC PDU can be either a data PDU or a control PDU. Data PDUs transfer higher layer data, and control PDUs relay acknowledgment info for RLC data PDUs.

The AM mode also handles HARQ reordering at the reception buffer in a similar way as the UM mode.

6.3.2 RLC Functions

As a summary, the following functions are supported by the RLC [6]:

- Transfer of upper layer PDUs;
- Error correction through ARQ (only in AM mode);
- Concatenation, segmentation, and reassembly of RLC SDUs (UM and AM);
- Resegmentation of RLC data PDUs (AM);

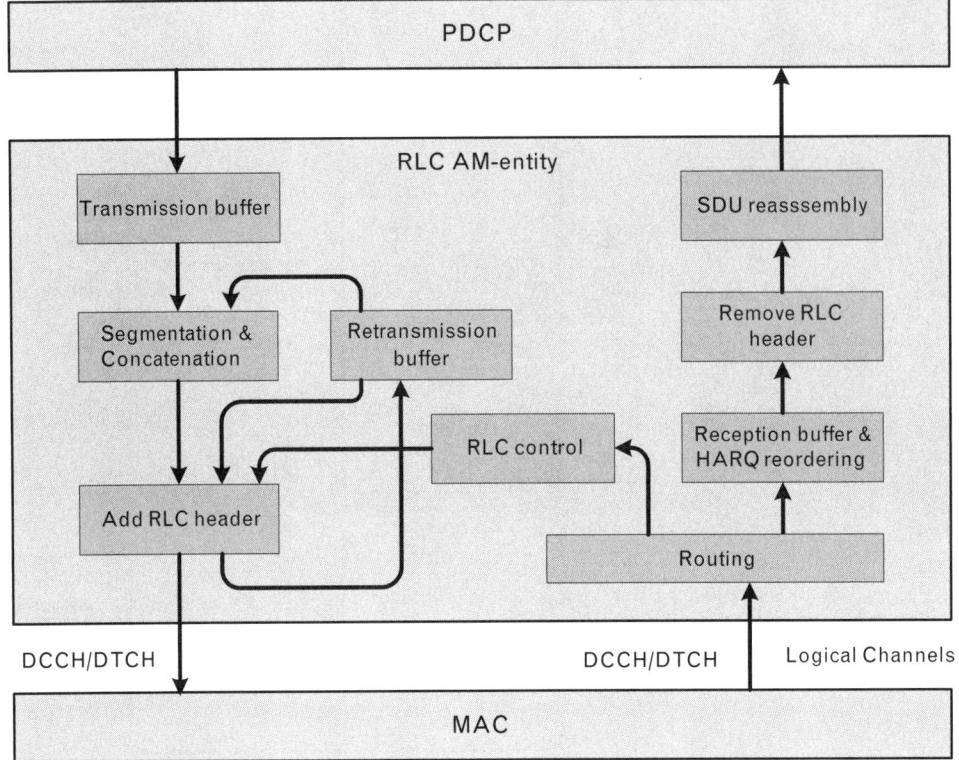

FIGURE 6.10 *The acknowledged mode entity in RLC.*

- Reordering of RLC data PDUs (UM & AM);
- Duplicate detection (UM & AM);
- RLC SDU discard (UM & AM);
- RLC re-establishment;
- Protocol error detection (AM).

The RRC layer is generally in control of the RLC configuration.

6.3.3 Logical Channels

Logical channels define what type of data is transferred. These channels define the data transfer services offered by the MAC layer; that is, the concept of logical channels is used in the interface between MAC and RLC layers. Instead of the uplink/downlink split, logical channels are typically classified by the type of data they carry (i.e., control and traffic channels) [2, 6].

6.3.3.1 Control Channels

These channels carry control plane information only. Note that control data in this case does not mean RLC layer control data, but higher layer control data that uses the control plane.

Broadcast Control Channel (BCCH)
This control channel provides system information to all mobile terminals that are within the coverage area of the eNodeB. It is mapped either to the BCH (in case of MIB system information) or to the DL-SCH (all other system information) transport channel. The BCCH is a downlink-only channel. Within the RLC layer, the BCCH uses the TM mode for its transmissions.

Paging Control Channel (PCCH)
PCCH relays paging information to all UEs within the cell coverage area. Its purpose is to inform the UE about the need to set up a connection. It is also used to notify UEs about system information changes on the BCCH. The PCCH is mapped to the PCH transport channel, and it is also a downlink-only channel. Within the RLC layer, the PCCH uses the TM mode for its transmissions.

Common Control Channel (CCCH)
CCCH is used to transfer control information when there is no established connection yet. In practice this means random access procedure messages that are used to establish such a connection. The CCCH is mapped to the UL-SCH or the DL-SCH, so it can be used in both directions: uplink and downlink. Within the RLC layer, the CCCH uses the TM mode for its transmissions.

Multicast Control Channel (MCCH)
This channel transmits control information that is used to support MBMS services. It is mapped either to the DL-SCH or to the MCH. DL-SCH mapping is used if this is a single-cell MBMS service, and MCH mapping in case the data requires MBSFN combining (i.e., the same data is broadcast over multiple cells). The MCCH is a downlink-only channel that uses the UM mode for its transmissions within the RLC layer.

Dedicated Control Channel (DCCH)
DCCH is the control channel that is used for user-specific control information. This implies that there has to be a connection between the UE and the eNodeB before the DCCH can be used. The DCCH carries all kinds of connected mode signaling data, related to procedures such as handovers, measurements, security, and so on. The DCCH is used both in the uplink

and the downlink. Within the RLC layer, the DCCH uses the AM mode for its transmissions.

6.3.3.2 Traffic Channels

These traffic channels carry user-plane data (i.e., content that is generated by some application). Two traffic channel types are defined for LTE: the DTCH for point-to-point connections and the MTCH for point-to-multipoint connections.

Dedicated Traffic Channel (DTCH)
DTCH is used for transferring user data over dedicated point-to-point connections. That is, the UE must have a connection with the eNodeB before this channel can be used. The DTCH can be used both in the uplink and in the downlink. Within the RLC layer, the DTCH can use either the UM or the AM mode for its transmissions.

Multicast Traffic Channel (MTCH)
MTCH is used for the transmission of multicast data (i.e., data that is sent over a point to multipoint connection). At the moment the only service that can employ the MTCH is MBMS. A multicast channel is obviously used only in the downlink, and it uses the UM mode within the RLC layer.

Figures 6.11 and 6.12 depict the mapping between logical channels, transport channels, and physical channels for uplink and downlink, respectively.

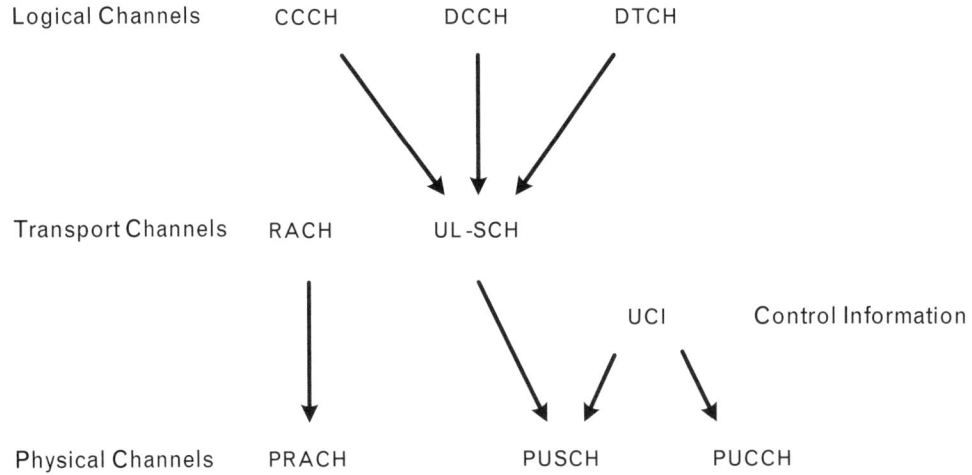

FIGURE 6.11 *Channel mapping in LTE air interface—uplink.*

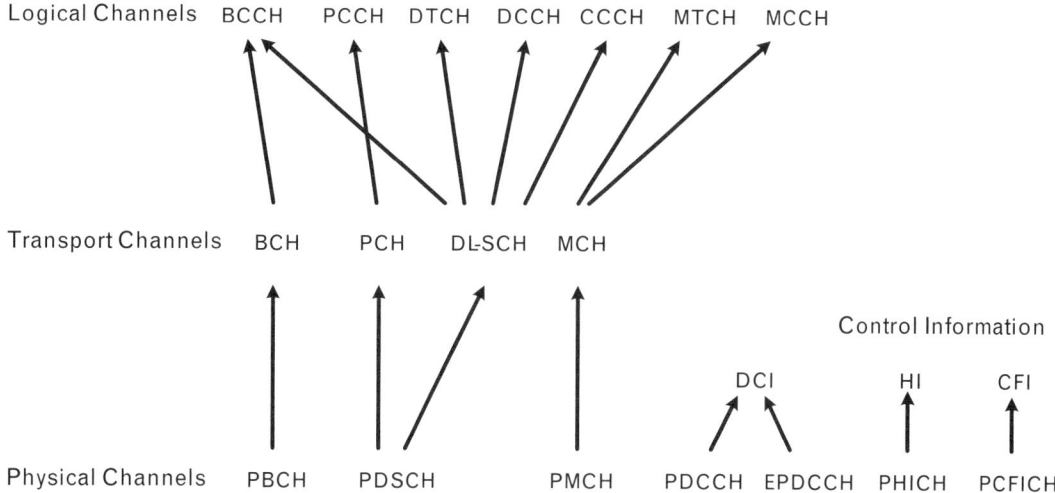

FIGURE 6.12 *Channel mapping in LTE air interface—downlink.*

6.4 Packet Data Convergence Protocol (PDCP)

The PDCP layer [7] has three main tasks:

- Header compression;
- Security functions;
- Handover support.

PDCP has different structure and functionality for user and control plane data. The user plane has more functionality (see Figure 6.13), whereas the control plane (see Figure 6.14) only implements the security functions.

Header compression is an important function that is needed with VoIP services. VoIP packets typically have very small payload, so much so that the VoIP header part is usually bigger in size than the payload. If every VoIP packet carries a full RTP/UDP/IP header, then the channel capacity is clearly wasted. The header compression algorithm used in LTE is the robust header compression (ROHC) protocol by IETF. The basic principle in ROHC is that the full header is sent only in the initialization phase, and after that only modified parameters are transmitted.

ROHC is efficient in reducing the header size: In the case of IPv4 the header size is 40 bytes, and in the case of IPv6 it is 60 bytes. Once the compression is done, the header size will be reduced to 4–6 bytes.

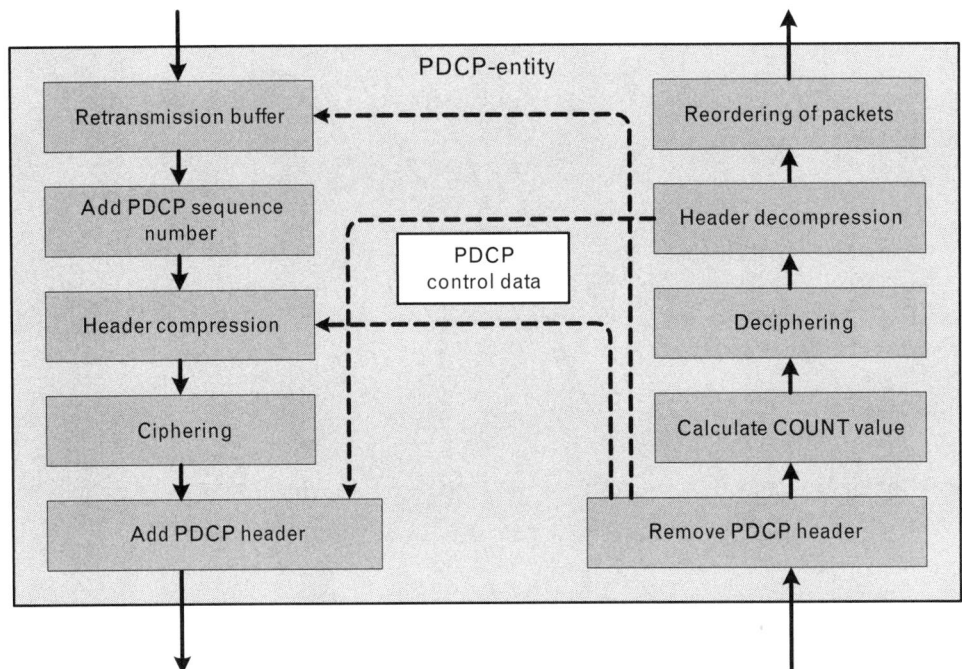

FIGURE 6.13 *PDCP layer, user plane.*

Header compression is performed only for the user plane, and only for VoIP packets. In fact, its implementation in the UE is not mandatory; only UEs with VoIP support need to have it.

Security in PDCP includes two functions: ciphering and integrity protection. Ciphering encrypts the data so that it cannot be read by intruders. Integrity protection verifies that the message at the receiver is the same as what the transmitter sent. Note that integrity protection itself does not prevent an intruder from reading the message; it just prevents its unauthorized modification. Therefore, in many cases both ciphering and integrity protection are used for control plane data. Ciphering is performed to both U-plane and C-plane data, integrity protection only to C-plane data. Ciphering and integrity protection are discussed in [8].

Handover support in PDCP context means support for user data handling during handovers. Depending on the type of data (i.e., whether the data transmission employs the UM or the AM mode in the RLC layer), the PDCP layer can perform a seamless or a lossless handover.

If the data transmission uses the UM mode, then delay is an important QoS parameter for this data. However, the loss of a few data packets is acceptable in the UM mode. For example, streaming data fits the description for such a service. In this kind of case the PDCP performs a seamless handover. This type of handover is designed to be as quick as possible. Complexity is kept at minimum to minimize the delays. However, as a result of

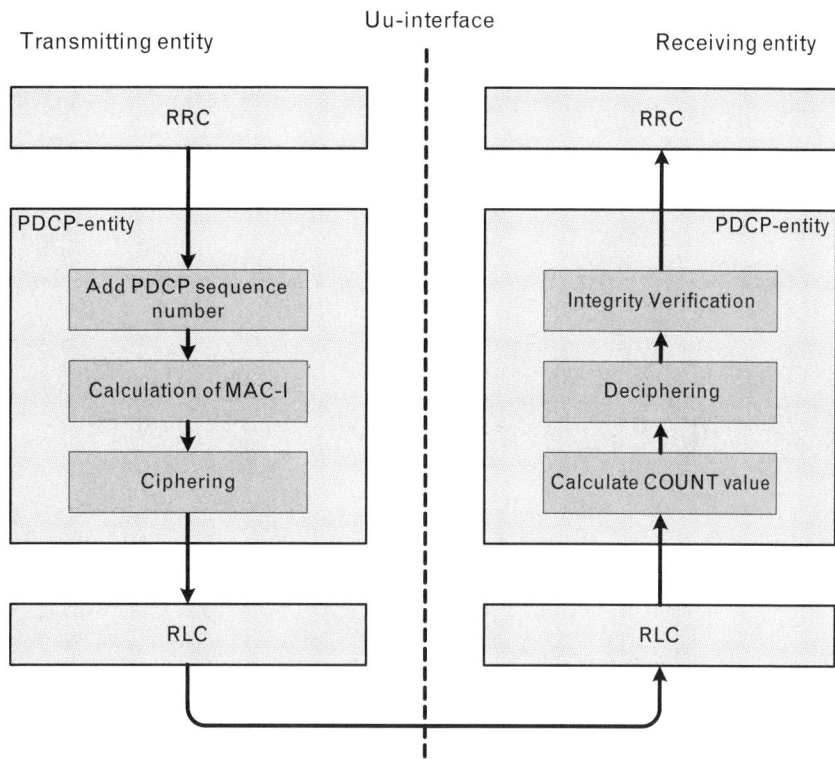

FIGURE 6.14
PDCP layer, control plane.

this, there are no checks at the receiver for missing data packets during the handover process, and therefore such PDCP SDUs that were already transmitted from the source entity, but not yet received at the target entity when the switchover took place, may have been lost. See [7] for details.

If the data transmission uses the AM mode, then it is obvious that lost data packets are not tolerated. However, small delays in the handover procedure will be acceptable since the AM mode will create them anyway. In this case a lossless handover is performed. This handover type requires more complex support mechanisms, and as a result longer delays will also take place. But the mechanisms introduced will guarantee that the data should be received without lost packets.

In a lossless handover the PDCP layer has to take a few special steps in order to guarantee that all data packets are delivered successfully:

1. The data packets already transmitted via the old link have to be retransmitted via the new radio link if no acknowledgment was received via the old link before the switchover took place.
2. Step 1 will result in duplicate data packets being received in cases where the original data packet got through, but its acknowledgment did not because the old radio link was cut before

acknowledgment(s) were delivered. Therefore, the receiver must check PDCP sequence numbers from received data packets and remove duplicates. To make this possible, PDCP sequence numbers must be maintained throughout the whole handover procedure.

3. The handover procedure also increases the possibility that data packets are received out of sequence at the receiver. Therefore, the receiver must use a data buffer for reordering the packets. Such a buffer exists in user-plane PDCP in any case, but its usage in a lossless handover requires that PDCP sequence numbers are maintained throughout the whole handover procedure.

6.5 Radio Resource Control (RRC)

6.5.1 Introduction

Radio resource control [9] is a big protocol task, no matter how you measure it. It has lots of functionality, its code size is large, and the RRC specification is also very large.

RRC exist in the control plane only; no user plane data is processed in the RRC layer. It configures and manages other air interface access stratum protocols. Other protocol tasks have their own control signaling, too, especially at the lower end of the protocol stack. Typically, however, RRC takes care of the initial configuration of protocol tasks and manages slowly changing parameters, whereas other tasks can self-configure rapidly changing parameters for which the configuration via RRC signaling approach would be too slow an alternative.

6.5.2 UE States

RRC functionality depends a lot on the UE state (sometimes also known as the RRC state). In LTE the UE state diagram is greatly simplified from the one in UMTS; now there are only two states: RRC_IDLE and RRC_CONNECTED. The transition between these states takes place when an RRC connection is established or released.

As readers may recall, in UMTS there were all together five UE states: UTRA_Idle, and four RRC_Connected substates: URA_PCH, CELL_PCH, CELL_FACH, and CELL_DCH. In LTE the substates have been removed (see Figure 6.15).

The UE states can be characterized as follows:

- RRC_IDLE:
  - DRX support (RRC configures the physical layer in the UE to monitor only the paging channel during certain periods). This is not

6.5 RADIO RESOURCE CONTROL (RRC)

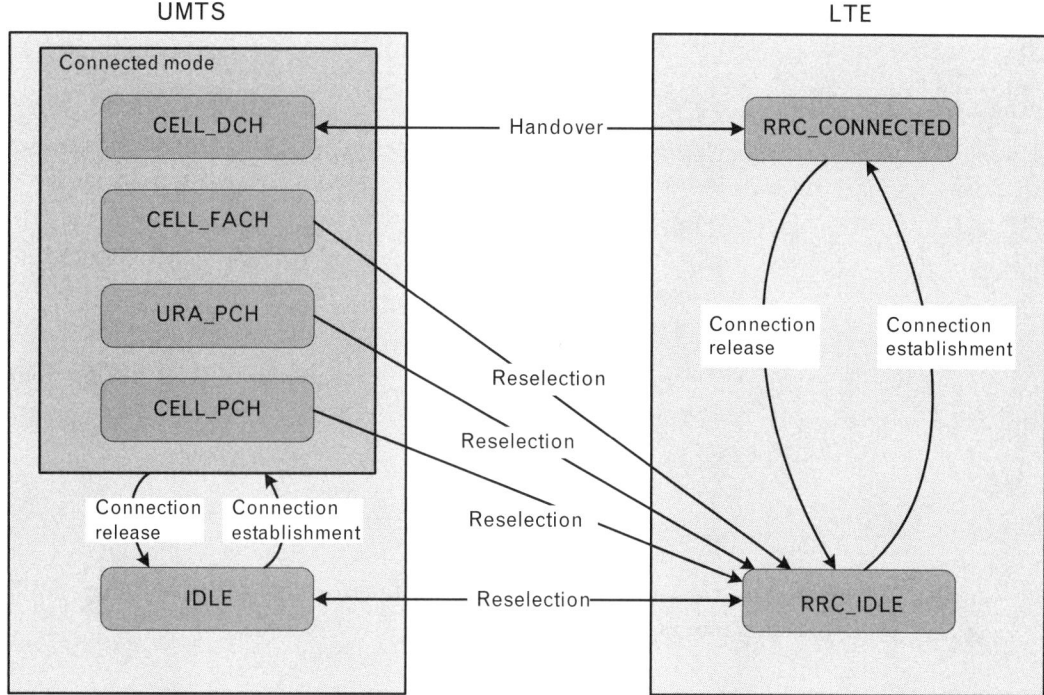

FIGURE 6.15 *RRC states in LTE and UMTS.*

a mandatory option, but a UE without DRX will quickly drain its batteries.

- UE controls its own mobility (i.e., the network does not stipulate which cell the UE should camp onto).
- In RRC_IDLE, the UE must:
  - Monitor a paging channel to receive notifications for incoming calls, system information changes, and PWS notifications (if supported).
  - Measure neighboring cells, and, based on these measurements, perform cell reselection;
  - Read system information.
  - Perform measurement logging (if supported).
- RRC_CONNECTED:
  - Data transfer to/from the UE.
  - DRX support.
  - Support for carrier aggregation (from Rel-10 onward).

- Network-controlled mobility—the network decides which cell the UE should use.
- The UE should:
  - Monitor the paging channel and/or SIB1 to receive notifications for system information changes and PWS notifications (if supported).
  - Monitor control channels associated with the shared data channel to receive data indications.
  - Provide channel quality and feedback information.
  - Perform neighboring cell measurements and measurement reporting.
  - Read system information.

6.5.3 RRC Functions

The list of functions the RRC has to perform is very long. They can be grouped into eight classes:

- Broadcast of system information;
- RRC connection control;
- Inter-RAT mobility;
- Measurement configuration and reporting;
- Generic protocol error handling;
- Support of self-configuration and self-optimization;
- Support of measurement logging and reporting;
- Other functions.

In the following sections we will briefly discuss the contents of each of these classes.

6.5.3.1 Broadcast of System Information

System information includes important configuration information about the network and the serving cell. Since this information is common to all users of the cell, it is broadcast to all UEs in the cell coverage area. System information is grouped into messages called the master information block (MIB) and a number of system information blocks (SIBs).

The MIB includes the most essential parameters that are needed for the UE to acquire other information from the cell: the carrier bandwidth and the current system frame number. The MIB is transmitted on BCH, which is a low-capacity channel, and thus the amount of data in MIB is kept at minimum. The system information in MIB is transmitted again every 40 ms; it has the shortest periodicity of any system information block.

System information blocks are numbered from 1 to 16. This is the status in Release 11—later releases can and will introduce additional SIBs once new features requiring configuration information are added to the system. LTE is designed to be easily extendable; new messages and message elements can be added to the system without problems in a backwards compatible way. SIB type 1 is transmitted in a message of its own; other SIBs are embedded into the system information (SI) message. One SI message can contain several SIBs—but only SIBs that have the same periodicity. SIB1 has an element called the schedulingInfoList, which indicates the presence of SIBs in each SI. Not all SIBs are transmitted in all networks or all cells. Some SIBs are feature-specific. If a network does not support a certain feature, then the corresponding SIB is not transmitted. Also, a UE that does not support a certain feature does not have to receive the related SIB. The SI and the SIBs are transmitted on the DL-SCH. The periodicity for SIB1 is 80 ms; for other SIBs the periodicity is configurable.

The information received in a SIB is valid until new and different information is received, which then overwrites the old info. The oncoming change in SIB contents is indicated with the parameter systemInfoModification in a paging message. If the contents are to be changed, then the UE has to receive SIBs again. If the UE misses this paging message (e.g., it is out of coverage temporarily), then it can check if SIBs have changed from SIB1, which carries a systemInfoValueTag parameter. If the value of this tag has changed since the UE last read it, then there has been a change in the contents of SIBs. Therefore a UE should always read SIB1 if it has been out of coverage even for a short period.

Note that not all system information parameter changes will result in a change in the systemInfoModification or in the systemInfoValueTag. Some SIBs contain rapidly changing information such as current timing info, which does not need to be read again after the initial SIB acquisition.

System information is broadcast information; the UE does not need to acknowledge its reception.

The list of SIBs in Release 11 is as follows.

Master Information Block

- Basic system parameters.

System Information Block Type 1

- Information needed for the UE to evaluate if it is allowed to access the cell.
- Scheduling of other system information blocks.

System Information Block Type 2

- Radio resource configuration information that is common for all UEs in the cell.

System Information Block Type 3

- Cell reselection information, mainly related to the serving cell.

System Information Block Type 4

- Neighbor cell–related information relevant only for intrafrequency cell reselection.
- Includes cells with specific re-selection parameters as well as blacklisted cells.

System Information Block Type 5

- Contains information about other E-UTRA frequencies and interfrequency neighboring cells relevant for cell reselection.

System Information Block Type 6

- Contains information about UTRA frequencies and UTRA neighboring cells relevant for cell reselection.

System Information Block Type 7

- Contains information about GERAN frequencies relevant for cell reselection.

6.5 RADIO RESOURCE CONTROL (RRC)

System Information Block Type 8

- Contains information about CDMA2000 frequencies and CDMA2000 neighboring cells relevant for cell reselection.

System Information Block Type 9

- Contains a home eNB (HeNB) name.

System Information Block Type 10

- Contains an ETWS primary notification.
- The ETWS primary notification is a short and quick warning of a dangerous situation.
- ETWS is the Japanese PWS system.

System Information Block Type 11

- Contains an ETWS secondary notification.
- The ETWS secondary notification is typically a longer message related to the threat indicated by the ETWS primary notification.
- The ETWS secondary notification may be segmented over several SIB11 messages.

System Information Block Type 12

- Contains a CMAS, a KPAS, or an EU-alert notification.
- CMAS is the North American PWS system.
- KPAS is the Korean PWS system.
- EU-alert is the EU PWS system.

System Information Block Type 13

- Contains MBMS-related information.

System Information Block Type 14

- Contains information about extended access barring for access control.

System Information Block Type 15

- Information related to mobility procedures for MBMS reception.

System Information Block Type 16

- Information related to GPS time and coordinated universal time (UTC).

6.5.3.2 RRC Connection Control

RRC connection control is a very broad topic, encompassing a long list of functionality starting from paging and ending with the release of a connection. It includes security management, handovers, carrier aggregation, QoS control, and so on. In fact, RRC connection control is a topic worthy of a book in itself; therefore, we can only scrape the surface when discussing this topic.

Note that many of the following functions are further discussed in Chapter 9, Procedures.

The list of functions belonging under RRC connection control banner include the following:

- Paging;
- RRC connection establishment;
- Security management;
- RRC connection reconfiguration;
- Counter check;
- RRC connection re-establishment;
- RRC connection release;
- Proximity indication.

The purpose of the paging procedure is as follows:

- To transmit paging information to a particular UE that is in RRC_IDLE state;

6.5 RADIO RESOURCE CONTROL (RRC)

- To inform all UEs about a system information change;
- To inform all UEs about a PWS notification.

RRC connection establishment is always initiated by the UE. It may follow a paging message reception, or it may be a result of a request from higher layers in the UE. Before launching an RRCConnectionRequest, the UE has to check from received SI that it is allowed to access the current cell. This message will trigger a random access procedure in the MAC layer, but note that random access is purely a MAC layer procedure in LTE.

Security management includes the management and generation of security keys. Security in the access stratum involves two functions: ciphering and integrity protection. In the access stratum, both of them are performed at the PDCP layer. However, RRC handles the security management and configures the required security parameters. Note that NAS has its own security functions.

The integrity protection algorithm is common for signaling radio bearers SRB1 and SRB2. The ciphering algorithm is common for all data bearers and control bearers SRB1 and SRB2. Neither integrity protection nor ciphering applies to SRB0. Both ciphering and integrity protection are always activated together. They can never be deactivated once activated, but it is possible to switch to a "NULL" ciphering algorithm. The null algorithm can be used, for example, when making an emergency call without a USIM.

The various security keys [8] required by the access stratum are all derived from a key called access security management entity (K_{ASME}). K_{ASME} is stored in the USIM in the UE, and in the authentication centre in the home subscriber server (HSS). The AS uses three different security keys:

- KRRCint for the integrity protection of RRC signaling;
- KRRCenc for the ciphering of RRC signaling;
- KUPenc for the ciphering of user data.

These keys are derived from the KeNB key, which is again based on the K_{ASME} key. Once a connection is established, a new set of AS keys is generated. This also applies to handovers and connection re-establishment procedures: a new set of keys will be needed then.

RRC connection reconfiguration is a multipurpose procedure. It can establish, modify, or release radio bearers; perform handovers; set up, modify, or release measurements (there is no separate "measurement command" in LTE); and add, modify, or release SCells in case of carrier aggregation. This procedure is always initiated by the network. It may include the reconfiguration of the physical, MAC, RLC, and PDCP layers, or a subset of those.

The counter check procedure is used by the network to request the UE to verify the amount of data sent and received on each data bearer. Both sides will maintain the counter value independently, and this procedure verifies that these values match. The purpose of this procedure is to detect packet insertion by an intruder (a "man in the middle attack").

When a failure occurs in the radio connection, the UE may attempt to re-establish the existing connection rather than to start from the beginning. The procedure for doing this is called the RRC connection re-establishment. Since the old cell obviously had problems, the UE must first find a new cell. If this is successful, the UE can send an RRC connection re-establishment request message, which will trigger a random access procedure in the MAC layer. This procedure, if successful, will re-establish the signaling connection only (SRB1). All other bearers will then have to be re-established with the RRC connection reconfiguration procedure.

The RRC connection release procedure releases the RRC connection, which includes the release of the established radio bearers as well as all radio resources. As a result of this procedure, the UE will enter the RRC_IDLE state. Whereas the RRC connection establishment is always initiated by the UE, its release is always initiated by the network side with a RRCConnectionRelease message. This message is unacknowledged. Note that the UE can itself initiate a connection release, but that is done at upper layers and that process will later result in this network-initiated RRC connection release.

Proximity indication is sent by the UE to indicate that the UE is entering or leaving the proximity of a closed user group (CSG) member cell (i.e., there is a Home eNodeB nearby). This information is useful for the eNodeB, because it may not otherwise know if the UE is close to a CSG cell. Once the proximity is known, the eNodeB may configure the UE to perform measurements on the frequency used by the CSG cell and take other actions related to a handover to the CSG cell [10].

6.5.3.3 Inter-RAT Mobility

Inter-RAT mobility [2, 9], in its widest sense, includes both idle and connected mode mobility. In the idle mode, the mobility is UE-controlled, whereas in the connected mode it is controlled by the network. A prerequisite for inter-RAT mobility is that the UE supports the radio technology of the other network.

In the idle mode the UE gets to know the potential inter-RAT cells via broadcast system information. SIB6, SIB7, and SIB8 contain information on UTRA, GERAN, and CDMA2000 frequencies and neighboring cells, respectively. With this information the UE can measure the inter-RAT cells to find out if there are suitable cells for reselection. Inter-RAT cell reselection is based on absolute priorities where the UE tries to camp on the highest priority RAT available. Priorities are given in the system information, and thus they are valid for all UEs in a cell. However, specific priorities per

UE can be signaled in the RRC connection release message. If the criteria for a cell reselection are fulfilled, the UE can simply camp on the new cell, and there is no need to inform the network about it, though a cell reselection may trigger a location update procedure.

In the connected mode, the inter-RAT mobility procedure is much more complex, including, for example, security activation and the transfer of RRC context information.

Mobility to LTE is very much like a normal intra-RAT LTE handover. When the handover from another RAT is initiated by the other RAT, the UE is configured to use LTE radio resources via messages in the other RAT, and then a RRCConnectionReconfiguration message is sent to the UE via LTE. This will be followed by the ciphering activation, unless that was already done by the previous RAT.

Mobility from LTE is a more varied topic. There are (at least) three ways to perform such a procedure:

- *Handover:* the MobilityFromEUTRACommand message includes radio resources that have been allocated for the UE in the target cell; this option applies to mobility from LTE to UTRA and GERAN.

- *Cell change order:* the MobilityFromEUTRACommand message may include target cell system information. The cell change order procedure is applicable only to mobility from LTE to GERAN.

- *Enhanced CS fallback to CDMA2000 1xRTT:* the MobilityFromEUTRACommand message includes radio resources that have been allocated for the UE in the target cell. The enhanced CS fallback to CDMA2000 1xRTT may be combined with concurrent handover or redirection to CDMA2000 HRPD.

In case of mobility to CDMA2000, the eNodeB in LTE decides when to move to the other RAT while the target RAT determines to which cell the UE shall move.

6.5.3.4 Measurement Configuration and Reporting

This section is about measurements performed in the connected mode. The UE reports measurement information based on the measurement configuration as provided by the E-UTRAN. This information is provided using the RRCConnectionReconfiguration message.

The UE can be requested to perform the following types of measurements:

- Intrafrequency measurements: measurements at the downlink carrier frequencies of the serving cells;

- Interfrequency measurements: measurements at frequencies that differ from the downlink carrier frequencies of the serving cells;
- Inter-RAT measurements (UTRA/GERAN/CDMA2000).

The measurement configuration in the RRCConnectionReconfiguration includes the following parameters [9]:

1. *Measurement objects:* The objects on which the UE should perform the measurements. These objects are usually carrier frequencies, possibly with a list of cell specific offsets and a list of "blacklisted" cells.
2. *Reporting configurations:* A list of reporting configurations where each reporting configuration consists of:
    - Reporting criterion: The criterion that triggers the UE to send a measurement report. This can either be a periodic or a single event description.
    - Reporting format: The quantities that the UE includes in the measurement report (i.e., what the UE is supposed to measure) and the associated information (e.g., the number of cells to report). The quantity parameters depends a lot on the system to be measured (i.e., LTE, UMTS, GERAN, or CDMA2000) and can be, for example, received signal code power (RSCP), reference signal received power (RSRP), reference signal received quality (RSRQ), or received signal strength indicator (RSSI). In case of LTE, this parameter to be measured is either RSRP or RSRQ.
3. *Measurement identities:* A list of measurement identities where each measurement identity links one measurement object with one reporting configuration. The measurement identity is used as a reference number in the measurement report.
4. *Quantity configurations:* The quantity configuration defines the filtering for each measurement.
5. *Measurement gaps:* Periods that the UE may use to perform measurements (and during these periods no uplink or downlink transmissions for this UE are scheduled).

The measurement procedures distinguish the following types of cells:

1. Serving cells;
2. Listed cells: these are cells listed within the measurement objects;
3. Detected cells: these are cells that are not listed within the measurement objects but are detected by the UE on carrier frequencies indicated by the measurement objects.

The rules on what to measure and where are as follows:

- For E-UTRA, the UE measures and reports on the serving cells, listed cells, and detected cells.

- For inter-RAT UTRA, the UE measures and reports on listed cells and optionally on cells that are within a range for which reporting is allowed by E-UTRAN.

- For inter-RAT GERAN, the UE measures and reports on detected cells.

- For inter-RAT CDMA2000, the UE measures and reports on listed cells.

As explained earlier, a measurement report can be event triggered or periodical. In event-triggered reporting, the report is sent once a certain event takes place. In periodic reporting, the triggering event is the corresponding timer expiry. The triggering event can also be a mixture of event-triggered and periodic reporting: the first report is only sent once an event occurs, but after that the following reports are sent periodically until the maximum number of reports has been sent or the event reporting criteria are no longer fulfilled. Note that for each event-triggered reporting criteria, both the entering condition and the leaving condition for measurement reporting has been defined.

The list of event-triggered reporting criteria in LTE include:

- Event A1: Serving cell becomes better than a threshold.

- Event A2: Serving cell becomes worse than a threshold.

- Event A3: Neighbor cell becomes better than the serving cell plus an offset.

- Event A4: Neighbor cell becomes better than a threshold.

- Event A5: Serving cell becomes worse than threshold1, and neighbor cell becomes better than threshold2.

- Event A6: Neighbor cell becomes better than the secondary serving cell plus an offset.

For inter-RAT measurements, the following event-triggered reporting criteria have been defined:

- Event B1: Inter-RAT neighbor cell becomes better than threshold.

- Event B2: Serving cell becomes worse than threshold1, and inter-RAT neighbor cell becomes better than threshold2.

Measurements are reported in a MeasurementReport message.

In the example in Figure 6.16, the UE measures the RSRP from both the serving cells and a neighboring cell. The serving cell is getting weaker, while the neighboring cells grow in strength. At point A, the neighboring cells becomes stronger than the serving cell, but this does not cause any actions yet because in reporting criteria event A3, the neighboring cell must be an offset value better than the serving cell before it is considered for a measurement report.

Once the neighboring cell becomes an offset value better than the serving cell at point B, the UE starts a timer (*time to trigger*). Timer time to trigger does not contain a fixed value, but the UE can scale it based on its own speed.

If the neighboring cell remains an offset value better than the serving cell for a time period indicated by a time-to-trigger parameter, then the UE will send a measurement report to the eNodeB (point C).

6.5.3.5 Generic Protocol Error Handling

The generic error handling applies to cases where the error handling is not explicitly specified elsewhere in specifications. This may include errors such as ASN.1 coding errors, missing mandatory message fields, a variable value out of range, or a field value not understood. Basically these are errors that should not happen but still somehow manage to happen. In most other

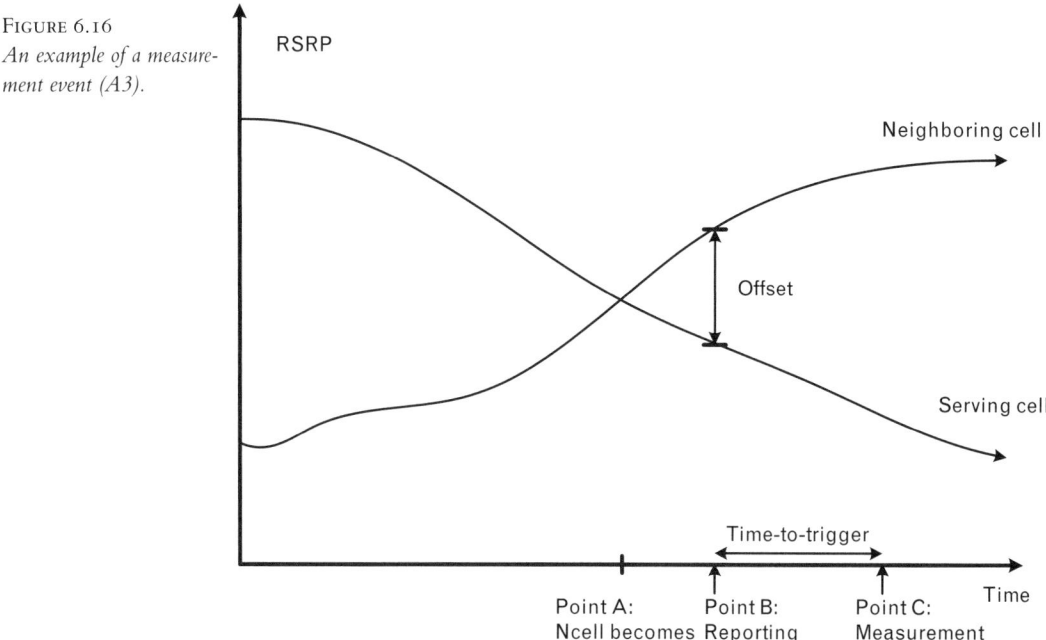

FIGURE 6.16
An example of a measurement event (A3).

cases, the error handling is already defined in relevant subclauses, but it is good to have generic error handling defined in case something unexpected happens.

6.5.3.6 Support of Self-Configuration and Self-Optimization

The self-optimizing network (SON) has become a popular catchword in LTE. SON is a new concept for LTE, not used in other mobile networks before LTE. SON includes various automated functions that can help the network operator to configure and autotune an LTE network. These functions can be very different in nature; for example, mobility load balancing (MLB) and energy saving do not have much in common—except that both of them can be used in improving the network performance, and they can be automated—and therefore they are part of SON.

SON is given a thorough treatment in Chapter 11. In this section we will just briefly mention that in some cases the RRC has to support various SON functions (e.g., by providing measurement results). It is worth noting that SON functions are typically radio access network (RAN) functions, residing in the eNodeB, which require little support from UEs over the air interface. SON mechanisms are also discussed in section 22 of [2].

6.5.3.7 Support of Measurement Logging and Reporting

This clause does not mean the measurements performed by the UE in the connected mode and defined in Section 6.5.3.4. Instead measurement logging refers to special measurements that the UE performs while in the idle mode (in some special cases also in the connected mode) in order to support network performance optimization. In LTE engineering jargon, this topic is known as minimization of drive tests (MDT) [9, 11], which describes its purpose well. Traditionally, when a network operator deploys a network, or adds eNodeBs to an existing network, the fine-tuning of the network is a major task. It involves lots of operator vans, driving around the coverage area of the network and measuring the radio conditions. The overall goal is to find, for example, coverage holes, areas with weak radio signals, areas with potential capacity problems, or areas where the neighboring cells can cause interference. This is an expensive undertaking and takes time if done properly. The idea behind MDT is to let the UEs to measure the radio conditions when they move around the network and report them back automatically.

These measurements are typically meant to take place in the idle mode, although there is a variant called the immediate MDT, which involves the UE to make the measurements in the connected mode. The measurements should have a time stamp and location information attached to them; otherwise, they are useless for MDT purposes. It is important that the amount of

MDT measurements per UE is kept within reasonable limits because they consume valuable battery power.

MDT measurements are initiated when the UE receives a LoggedMeasurementConfiguration message. The measurement quantity is fixed for the logged MDT (i.e., it is not configurable) and consists of both RSRP and RSRQ for LTE, both RSCP and Ec/No for UTRA, P-CCPCH RSCP for UTRA 1.28 TDD, Rxlev for GERAN, and pilot Pn phase and pilot strength for CDMA2000 (in all cases it is assumed that the serving cell is E-UTRAN).

The UE should perform the required measurements and store the results. It should not send them to the network by its own initiative; rather, it should indicate their presence with a flag in the RRCConnectionSetupComplete message (i.e., the next time the UE enters the connected mode). The network can then explicitly request the logged results with a UE information procedure (see Figure 6.17). The response message, UEInformationResponse, will contain the logged measurements.

6.5.3.8 Other Functions

Other functions include, for example, the transfer of dedicated NAS information and non-3GPP dedicated information, transfer of UE radio access

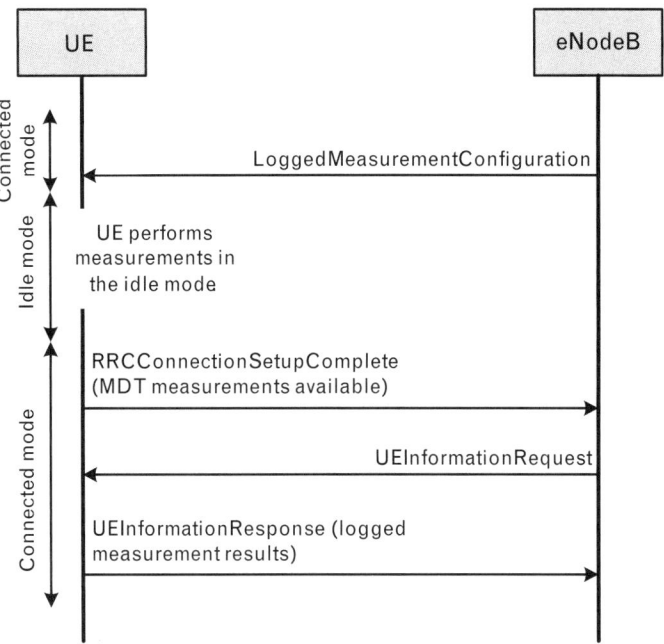

FIGURE 6.17
Minimization of drive tests procedure.

capability information, support for E-UTRAN sharing (multiple PLMN identities), and transfer of UE assistance information.

References

[1] 3GPP TS 36.401, v 11.2.0, Evolved Universal Terrestrial Radio Access Network (E-UTRAN); Architecture Description, 09/2013.

[2] 3GPP TS 36.300, v 11.7.0, Evolved Universal Terrestrial Radio Access (E-UTRA) and Evolved Universal Terrestrial Radio Access Network (E-UTRAN); Overall Description; Stage 2; 09/2013.

[3] 3GPP TS 36.321, v 11.3.0, Evolved Universal Terrestrial Radio Access (E-UTRA); Medium Access Control (MAC) Protocol Specification; 06/2013.

[4] 3GPP TS 36.216, v 11.0.0, Evolved Universal Terrestrial Radio Access (E-UTRA); Physical Layer for Relaying Operation; 09/2012.

[5] 3GPP TS 36.212, v 11.3.0, Evolved Universal Terrestrial Radio Access (E-UTRA); Multiplexing and Channel Coding, 06/2013.

[6] 3GPP TS 36.322, v 11.0.0, Evolved Universal Terrestrial Radio Access (E-UTRA); Radio Link Control (RLC) Protocol Specification; 09/2012.

[7] 3GPP TS 36.323, v 11.2.0, Evolved Universal Terrestrial Radio Access (E-UTRA); Packet Data Convergence Protocol (PDCP) Specification; 03/2013.

[8] 3GPP TS 33.401, v 12.9.0, 3GPP System Architecture Evolution (SAE); Security Architecture; 09/2013.

[9] 3GPP TS 36.331, v 11.5.0, Evolved Universal Terrestrial Radio Access (E-UTRA); Radio Resource Control (RRC); Protocol Specification; 09/2013.

[10] 3GPP TS 36.304, v 11.5.0, Evolved Universal Terrestrial Radio Access (E-UTRA); User Equipment (UE) Procedures in Idle Mode; 09/2013.

[11] 3GPP TS 37.320, v 11.3.0, Universal Terrestrial Radio Access (UTRA) and Evolved Universal Terrestrial Radio Access (E-UTRA); Radio Measurement Collection for Minimization of Drive Tests (MDT); Overall Description; Stage 2; 03/2013.

CHAPTER 7

Radio Access Network

7.1 Introduction

The radio access network (RAN) in LTE is called E-UTRAN, or evolved universal terrestrial radio access network. It is a bit unfortunate that the RAN in LTE is called evolved UTRAN, whereas in UMTS it is UTRAN. This gives the impression that the 4G technology employed is just a slightly upgraded version of 3G. However, as shown earlier, the air interface especially is based on completely different technologies, and the RAN architecture is also very different.

The E-UTRAN has a greatly simplified architecture when compared to UMTS UTRAN. In UTRAN the radio network controller (RNC) and the base station (NodeB) were separate entities, but in EUTRAN they have been combined into an evolved NodeB (eNodeB, or eNB).

Since we have already discussed the air interface in previous chapters, this chapter will concentrate on the network side of RAN, namely, the eNodeB, its variants, and the interfaces of RAN.

7.2 E-UTRAN Architecture

The E-UTRAN architecture [1, 2] is depicted in Figure 7.1. As seen, it has fewer entities than in the UTRAN architecture. The radio network controller and the base station have been combined into a single entity, eNodeB. This solution was chosen because it allows for shorter transmission delays for the signaling between the UE and the network.

The eNodeB connects to the core network, evolved packet core (EPC) via the S1 interface [3]. There are two variants of this interface:

- S1-AP interface exists between the eNodeB and the mobility management entity (MME).

- S1-U interface exists between the eNodeB and the serving gateway (S-GW).

126   RADIO ACCESS NETWORK

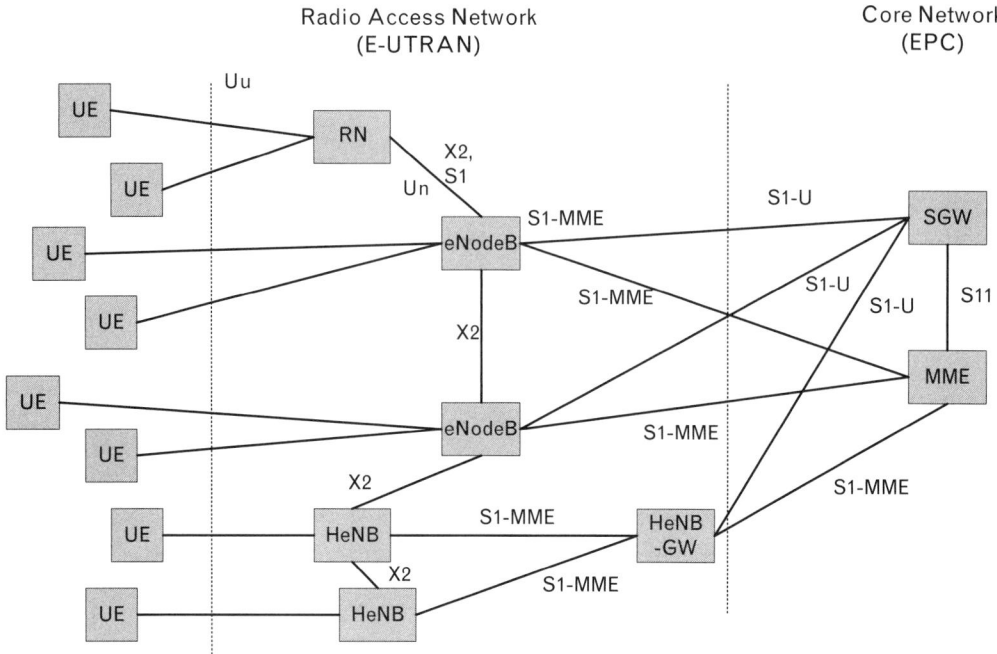

FIGURE 7.1 *E-UTRAN architecture.*

eNodeBs can also be connected directly to each other. This logical interface between them is called the X2 interface [4].

In addition, the E-UTRAN architecture includes home eNodeBs (HeNBs) [2, 5], which are special eNodeBs, designed for deployment in a private environment (e.g., in homes or offices). Logically, HeNB is a full eNodeB that connects to the core network via the S1 interface, and to UEs via the Uu interface. A HeNB is designed to be an affordably priced, easily deployable (without mobile operator's involvement) home electronics device that may or may not support a closed user group (CSG).

The home eNodeB gateway (HeNB-GW) is an optional gateway through which the home eNodeB accesses the core network. Its UTRAN counterpart, HNB-GW, is a mandatory element in UTRAN networks. However, even though the HeNB-GW is optional in EUTRAN, network operators typically have a HeNB-GW to optimize the operations and management of HeNBs. Otherwise, a very large number of HeNBs would be directly connected to the MME and the SGW. It is also possible to use this gateway only between the HeNB and the MME, and to have a direct interface between the HeNB and the S-GW.

The E-UTRAN may also include relay nodes (RNs) [2, 6]. A RN is a low-power base station that can be used to increase or improve the coverage of a LTE network. A RN is connected to a "host" eNodeB via a radio interface. The host eNodeB is called the donor eNodeB (DeNB). The RN is "architecturally transparent" in the sense that toward eNodeB it looks like a UE, and toward the UE it looks like a normal eNodeB. Relays are further discussed in Chapter 11.

7.3 eNodeB

7.3.1 eNodeB Introduction

The eNodeB combines the functionality of a base station and a radio network controller. In the past with 2G and 3G, the design aim was to keep the base station as simple and cheap as possible and to concentrate the intelligence in RAN into a radio network controller. However, since shorter transmission delays are an important design factor in LTE, the radio network controller was removed and its intelligence was moved into eNodeBs (i.e., closer to UEs).

This change has brought many advantages, the most important of which is that the network response time as experienced by UEs is much shorter than before. But there are also some problems. Because the radio network does not have a central "master" node anymore, the intelligence in the RAN is distributed. Therefore it is not obvious which node makes the decisions in procedures and functions that require inter-eNodeB interactions. For example, many new SON features, such as mobility load balancing, require careful design and distributed algorithms to work properly.

eNodeB protocol stacks are depicted in Figures 7.2 and 7.3. We have already discussed the air interface protocols in the previous chapter. The interface between the eNodeB and the core network, the S1 interface, comes in two different versions: S1-MME and S1-U. The S1-MME is in the control plane and the S1-U interface exists in the user plane.

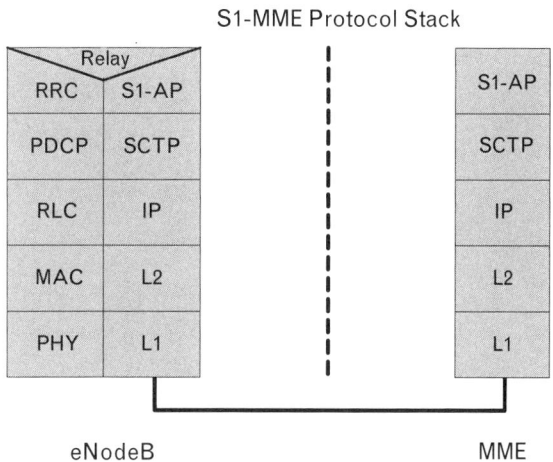

FIGURE 7.2
S1 protocol stack—control plane.

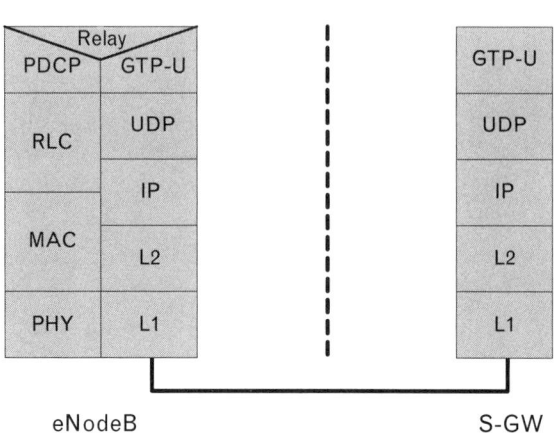

FIGURE 7.3
S1 protocol stack—user plane.

Between two eNodeBs exists the X2 interface. From a logical standpoint, the X2 is a point-to-point interface between two eNodeBs. However, as this is a logical interface, there is no direct physical connection between two eNodeBs. Figures 7.4 and 7.5 show X2 interface protocol stacks.

7.3.2 eNodeB Functionality

eNodeB has several important functions [2]:

- Radio resource management functions;
- Measurement and measurement reporting configuration;
- Access stratum security;
- IP header compression and encryption of user data;
- Selection of an MME at UE attachment;
- Routing of user plane data toward the serving gateway;
- Scheduling and transmission of paging messages;
- Scheduling and transmission of broadcast information;
- Scheduling and transmission of PWS messages.

Each of these functions is discussed briefly in following sections. It must be stressed here that a network operator is free to implement these functions as it sees best. In most cases the specifications do not restrict the selection and the implementation of the relevant algorithms. There are, of course, exceptions such as security. Security algorithms are strictly standardized, and their implementation must conform to the corresponding specifications.

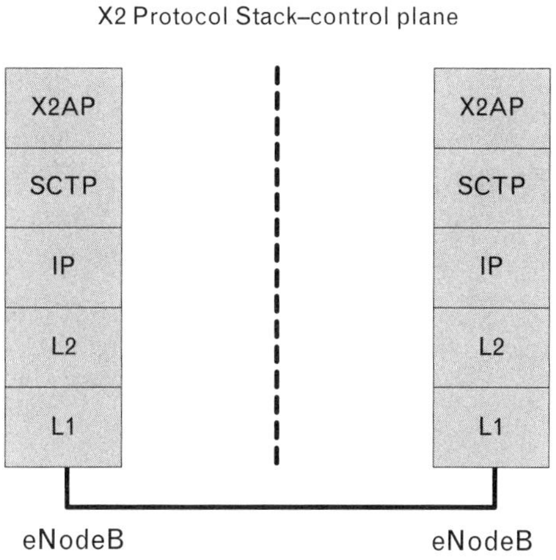

FIGURE 7.4
X2 protocol stack—control plane.

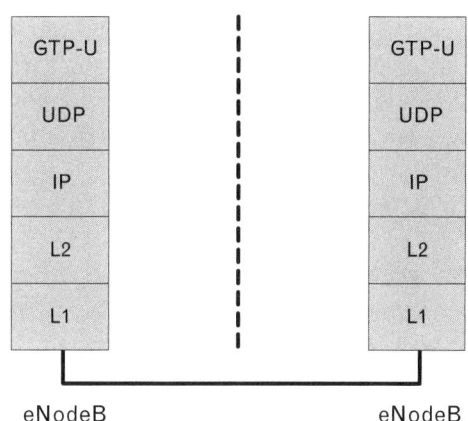

FIGURE 7.5
X2 protocol stack—control plane.

And even though the implementation of these functions is not guided by standards, there are detailed test specifications that will check that the performance of implemented functions is good enough. For example, the requirements for support of radio resource management can be found in [7].

Radio resource management (RRM) ensures the efficient use of available radio resources. It has several functions in its toolbox to achieve this:

- Radio bearer control (RBC);
- Radio admission control (RAC);
- Connection mobility control (CMC);
- Dynamic resource allocation (DRA);
- Intercell interference coordination (ICIC);

- Load balancing (LB).

Radio bearer control (RBC) decides what kind of radio bearers are assigned for each user and when. When setting up a radio bearer for a service, the RBC takes into account the overall resource situation in the E-UTRAN, the QoS requirements of in-progress sessions, and the QoS requirement for the new service. RBC also monitors the situation throughout the lifetime of the connection and can modify the bearer parameters due to mobility or other reasons. At the end of the connection, the RBC is involved in the release of radio resources associated with radio bearers.

Radio admission control (RAC) acts as a gatekeeper for the new radio bearer establishment requests. The goal of RAC is to provide high radio resource utilization by accepting new radio bearers if required radio resources are available, but also at the same time to ensure that existing sessions can be provided with high enough QoS. The latter requirement means that sometimes RAC has to reject new radio bearer requests. In order to do this, RAC must consider the overall resource situation in E-UTRAN, the QoS requirements, the priority levels and the provided QoS of existing sessions, and the QoS requirement of the new radio bearer request.

Connection mobility control (CMC) manages the radio resource reallocation that is due to UE mobility. This applies to both the idle and the connected mode mobility. In the idle mode, the UE makes the cell reselection decision autonomously, but CMC can control the reselection algorithm by setting the relevant parameters such as threshold and hysteresis values. These parameters are broadcast to all UEs in the cell. In the connected mode, handovers are controlled by the eNodeB. There is a long list of parameters the CMC can take into account when making handover decisions, such as radio measurements, neighbor cell load, traffic distribution, and transport and hardware resources.

Dynamic resource allocation (DRA) is also known as packet scheduling (PS). The task of DRA is to allocate resources to UEs both in the uplink and in the downlink. These resources are not only channel resources but also memory resources (packet buffers) and data processing resources. DRA is very dynamic as its name suggests: it may be necessary to reallocate the resources to UEs in every subframe. As with other RRM functions, the DRA can take a long list of measured parameters as inputs to its scheduling algorithm. These include at least the QoS requirements of radio bearers, the channel quality information, packet buffer status, and interference.

Intercell interference coordination (ICIC) has to manage radio resources in a way that minimizes the intercell interference. Intercell interference may be a serious problem for users near the cell edges and in pico/femto cells that are located inside macro cells but are operating on the same frequency. ICIC is a multicell function that has to take into account parameters from several neighboring cells. This makes it difficult to design an efficient algorithm, since there is no central radio controller node in E-UTRAN and all eNodeBs are equal. Therefore some degree of configuration by operation and maintenance (O&M) may be necessary in order to achieve a satisfactory ICIC solution. ICIC includes a frequency domain component and a time domain component. The frequency domain ICIC aims to manage radio resource blocks in a way that minimizes interference in the frequency domain. This requires coordination on the use of frequency resources across multiple cells. The time domain ICIC introduces a concept called *almost blank subframe* (ABS). ABSs are subframes that are transmitted with greatly reduced transmit power. Only the most essential control channels are transmitted during ABS periods. The ABS patterns are signaled to the eNodeB of the victim cell, and it can schedule the UEs suffering from interference to transmit their own data during the ABS frames only. The time domain ICIC is especially designed to deal with interference caused by a macrocell toward a pico/femto cell. The obvious drawback with ICIC schemes is that most of them tend of reduce the overall throughput of the cell because not all channel resources are used. The preferred ICIC method may be different in the uplink and in the downlink.

Load balancing (LB) aims to distribute the data traffic across multiple cells in a way that results in more evenly loaded cells. The LB algorithm tries to find such UEs that can be handed over from highly loaded cells to underutilized cells. The QoS of existing connections should be maintained even after such handovers. In most cases, load balancing is a difficult task because typically only UEs near cells edges can be considered candidates for load balancing, unless the network has more than one frequency channel in its use—in which case, the LB algorithm can perform interfrequency handovers to ease the congestion on the heavily loaded cell.

Measurement and Measurement Reporting Configuration
This topic was already discussed in Section 6.5.3.4. The eNodeB has to decide what kind of measurements and when are needed from each UE. As already seen in previous sections, the eNodeB needs measurement information for a wide variety of management functions. However, it should not overburden the UE with unnecessary measurement requests, since they do consume power.

Access Stratum Security
Access stratum (AS) and nonaccess stratum (NAS) security procedures and keys are kept separate in LTE. In AS, the security issues are handled by the eNodeB; in NAS the main actor is MME.

It is worth noting that even though both AS and NAS level security keys are derived from the same "root" key K_{ASME}, it is not possible to determine NAS level keys from AS keys or vice versa.

The same SecurityModeCommand is used to activate both ciphering and integrity protection simultaneously, if both are to be employed. It is not possible to launch one process first and then another later. However, AS and NAS use separate commands to initiate security procedures and may do those at different times.

A sequence number, COUNT, is used as an input to ciphering and integrity protection procedures. The same COUNT number must only be used once for a given data packet on the same radio bearer in the same direction, except when sending an identical retransmission. However, the same COUNT value can be used for both ciphering and integrity protection.

Security is a very wide topic, worthy of a book in itself. This book can only discuss the basic security principles in LTE. For more information, see [2, 8, 9].

IP Header Compression and Encryption of User Data
User data is transmitted in IP packets (either IPv4 or IPv6) that have a sizeable header. Normally, this is not a big problem, but with VoIP applications the amount of data in each packet is typically very small, and the header part can be larger than the data part. Therefore IP header compression can be performed for IP headers in case of VoIP. The header compression algorithm used in LTE is the robust header compression (ROHC) protocol [10]. The basic principle in ROHC is that the full IP header is sent only in the initialization phase, and afterward only the modified parameters are transmitted. Header compression is performed for only the user plane, and currently only for VoIP packets. Its implementation in the UE is not mandatory; only UEs with VoIP support need to have it. The eNodeB is in charge of when and which UEs should support the IP header compression. The compression/decompression function itself is a PDCP layer function.

The reason the PDCP has two different functions for security is that ciphering and integrity protection perform different, but complementary, tasks. Ciphering encrypts the data so that it cannot be read by intruders. Integrity protection verifies that the message at the receiver is the same as what the transmitter sent. Note that integrity protection itself does not prevent an intruder from reading the message; it just prevents its unauthorized modification. Therefore in many cases both ciphering and integrity protection are used for control plane data. Ciphering and integrity protection are discussed in [8].

Ciphering, deciphering, and integrity protection are performed at the PDCP layer. Once a connection is established, the eNodeB has to calculate an access stratum base key K_{eNB}, based on the K_{ASME} key [8]. This key is used to derive three further keys that are used in ciphering and integrity protection. The eNodeB activates the access stratum security with a SecurityModeCommand at the RRC layer. Ciphering is performed to both U-plane and C-plane data, integrity protection only to C-plane data.

Selection of an MME at UE Attachment
Normally, when a UE initiates a connection, it will send a globally unique MME identifier (GUMMEI) to the eNodeB in RRCConnectionSetupComplete message. This parameter identifies the MME the UE was last registered with. It will then be used by the eNodeB to select the correct MME from a pool of available MMEs. However, GUMMEI is not always known to the UE, or it may be that the eNodeB does not have access to the MME identified. In that case the eNodeB has to select the serving MME by other criteria, provided that there are more than one MME available for this eNodeB. In this selection, the eNodeB can employ, for example, some load sharing algorithms.

Routing of User Plane Data Toward the Serving Gateway
The selection of the S1-U interface for a given data connection is done within the EPC and signaled to the eNodeB by the MME.

Scheduling and Transmission of Paging Messages
The MME initiates the paging procedure by sending a paging message to the eNodeB. The eNodeB will then page the UE in cells that belong to tracking areas (TA) listed in the paging message. Note that each cell can belong to only one TA, but one eNodeB can contain cells belonging to different TAs.

Paging messages are also used to inform UEs about an oncoming change in system information. Also, they will warn the UEs about new PWS messages being sent in SIB10, SIB11, and SIB12.

Scheduling and Transmission of Broadcast Information
The MIB is carried on the BCH. All other SI messages are carried on the DL-SCH, where they can be identified through the SI-RNTI (system information RNTI). SI-RNTI is an identifier, transmitted on the PDCCH, indicating that there is system information being transmitted in the associated DL-SCH. Both the MIB and SystemInformationBlockType1 use a fixed schedule, with a periodicity of 40 and 80 ms, respectively, while the scheduling of other SI messages is flexible and indicated by SystemInformationBlockType1. It is up to the eNodeB to decide the periodicity of individual SI messages.

The eNB may schedule DL-SCH transmissions containing logical channels other than BCCH in the same subframe used for BCCH. System information may also be provided to the UE by means of dedicated signaling (e.g., during the handover).

Scheduling and Transmission of PWS Messages
PWS messages originate from the cell broadcast center (CBC). These messages are forwarded to eNodeBs by the MME. PWS messages are broadcast

on SIB10, SIB11, and SIB12 messages. The schedule information for the broadcast is received along with the broadcast message content from the CBC. Because of the nature of these warnings, it is imperative that they are delivered quickly. Typically, these messages are also sent repeatedly to ensure their successful reception. There are mechanisms in place that ensure that only one copy of the same message is displayed to the user. That is, the user knows that every message displayed is a new warning message that requires special attention. In case of ETWS messages, the secondary notification is not as urgent as the primary notification, and indeed it may be segmented and scheduled over several SIB11 messages.

The eNodeB is also responsible for paging the UE to provide an indication that a warning notification is being broadcast.

PWS handling is specified in [2, 11, 12].

7.4 Home eNodeB

7.4.1 Introduction

Home eNodeB (HeNB) is a small low-powered LTE base station designed to be used in offices or homes (i.e., it is deployed by someone else than the mobile network operator). In marketing talk, HeNBs are commonly known as LTE femto cells, since the HeNB is an engineering term only. Note that another popular term for micro cells, a pico cell, refers to operator deployed small cells.

The HeNB, in principle, has all the functionality of a full macro eNodeB. It is connected to the evolved packet core (EPC) via a fixed connection, typically the Internet, and to UEs via normal air interface. One notable difference between the eNodeB and HeNB is that whereas an eNodeB can manage several cells, a HeNB controls only one cell. The HeNB system logical architecture is shown in Figure 7.6.

HeNBs can have an X2 interface between them, just like "real" eNodeBs. Toward the evolved packet core (EPC) they connect through the standard S1 interface. It is also possible to deploy a HeNB gateway between the HeNB and the EPC; it would act mainly as a signal traffic concentrator.

A HeNB can operate in three different access modes:

- *Closed access mode:* HeNB provides services only to the members of its closed subscriber group (CSG).

- *Hybrid access mode:* HeNB provides services both to its associated CSG members and to nonmembers, but CSG members are prioritized over nonmembers.

FIGURE 7.6
HeNB architecture.

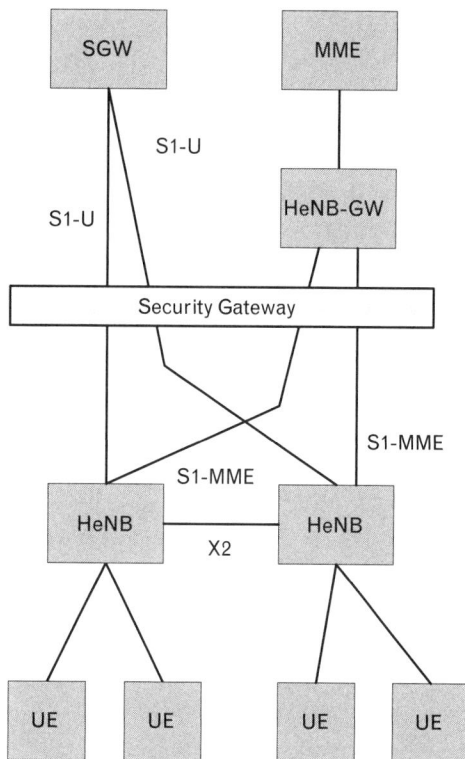

- *Open access mode:* HeNB appears as a normal eNB. Anybody can access its services, as long as they have a subscription or a roaming agreement with the mobile operator.

A HeNB will have a connection to mobile operator's O&M system, and the operator can carry out O&M functions on the HeNB. These can include remote configuration of the HeNB, software upgrades, detecting and reporting changes in RF conditions, and performing general O&M tasks. One major worry for the operator is the interference caused by HeNBs. Even though the operator can configure the HeNB via the O&M connection, it will not (in most cases) physically install the HeNB. The owner of HeNB may deploy it as he wishes, and as a result the location may not be optimal from the whole network point of view; that is, there is no network planning. For example, the HeNB may transmit a radio beam through a window, which will then cause interference to mobile users who happen to be within the beam footprint. The problem may be even worse if the HeNB operates in the closed access mode, because with the open access mode the users suffering from HeNB interference would have the option to make a handover to the HeNB. However, in most cases the interference problems can be avoided if the HeNB implements interference mitigation techniques. Of course, interference can happen both ways: interference caused by the macro cell can result in serious problems in HeNBs, especially if the HeNB is close to the macro cell transmitter. One approach to solve such a problem is to employ almost blank subframes (ABS) in the macro cell. This solution was already discussed in Section 7.2.1.2.

The easiest solution for HeNB interference problems would be to deploy all HeNBs on a separate operating frequency. Unfortunately this is not always possible since radio spectrum is expensive and there is always a shortage of it (at least if you ask network operators).

In some countries there is a requirement that it should be possible to locate the caller who has made an emergency call. This means that the exact position of the HeNB should be known to the network operator. The network operator can ask the HeNB location when the HeNB is registered with the operator for the first time, but the problem is that there is no guarantee that the device will stay in that location. For example, the HeNB owner may move to a new address with the HeNB and forget to update the address register. Or she may decide the take the HeNB with her to her holiday cottage for a weekend. A GPS device can be added to a HeNB to provide its location, but this approach may not work properly in all cases since HeNBs are typically installed indoors where the GPS coverage is poor.

The ability to locate the HeNB is also important for reasons other than emergency calls: it is easy to move a small HeNB around, so much so that there may be a temptation to take it to another country. When traveling, for example, it can be connected to a normal broadband connection, and at least in theory it could be used to make cheap calls back to home country. There are two problems with this approach:

1. The home PLMN operator most probably does not have a mobile network operating license in the other country.
2. A HeNB uses licensed spectrum, and it is tuned to use a specific part of that spectrum. In another country, this spectrum is licensed to some other operator or to some other purpose altogether.

HeNB service requirements are defined in [5]. Other useful HeNB information can be found from [2].

7.4.2 Closed Subscriber Group (CSG)

CSG refers to a defined group of users that are allowed to access a HeNB that operates in a closed access mode. Typically, the UE's identities belonging to a CSG are registered when the HeNB is first set up, but it is also possible to modify the CSG list later.

A closed access mode HeNB allows connections only from a UE that is a member of the corresponding CSG. Most HeNBs are of this type, and it is understandable that a HeNB owner does not want the whole neighborhood to use his broadband connection and to fill it with data traffic. This would certainly be the case if a HeNB were to operate in an open access mode in a densely built area. A hybrid access mode HeNB also allows all UEs to access it, but CSG members are given higher priority in case of congestion.

7.4.3 HeNB Mobility

Mobility between HeNBs and eNodeBs is supported both ways: to and from the HeNB. From Release 9 onward mobility between HeNBs is also supported (using the S1 interface), and from Release 10 onward the HeNB to HeNB handover can be done via the X2 interface. However, the direct X2 handover is allowed only if no access control by the MME is required. That is, both HeNBs should operate in open or hybrid access mode, or their CSG IDs should be the same.

Mobility from a HeNB to a eNodeB follows the standard intra-RAN handover procedure, but mobility to a HeNB requires some special attention. If the target HeNB is operating in the closed access mode, the MME has to check that the UE is a member of the CSG. In case of hybrid access mode, the MME accepts the handover even if the UE is not a member of the CSG, but the nonmember status is indicated to the target HeNB, which may "de-prioritize" this UE in case of congestion.

7.4.4 HeNB Gateway

The E-UTRAN architecture may also employ an optional home eNB gateway (HeNB GW) to allow the S1 interface between the HeNB and the EPC to support a large number of HeNBs in a scalable manner. However, it is likely that this gateway is used only in the S1-MME interface. The U-plane (S1-U interface) will most often have a direct connection to the S-GW. The HeNB GW is located in the operator's premises; it is not something a home user would purchase. A typical HeNB GW can serve tens of thousands of HeNBs.

The HeNB GW appears to the MME as an eNB, and to the HeNB it appears as an MME. Therefore the HeNB GW is "transparent" in the sense that neither entity, HeNB or MME, needs to know about its existence. The S1 interface between the HeNB and the EPC stays the same, regardless of whether the HeNB is connected to the core network via a HeNB GW or via a direct interface. See Figures 7.7 and 7.8 for HeNB GW protocol stacks.

7.4.5 Traffic Offloading

Traffic offloading refers to technologies that route data traffic from the U-plane directly to the Internet or to some other IP network, bypassing the

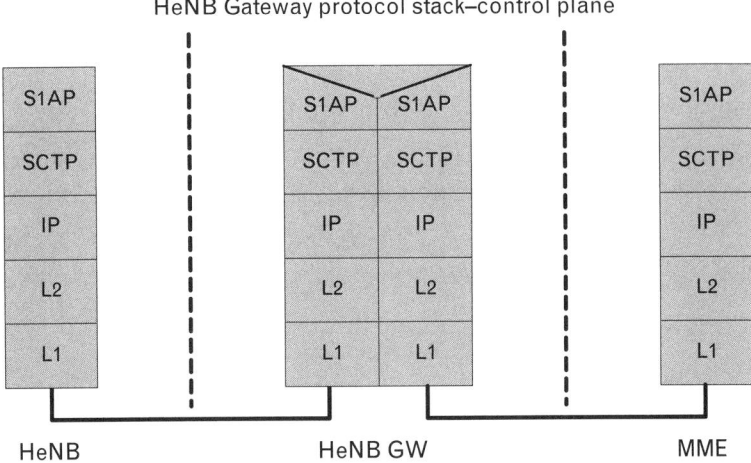

FIGURE 7.7
HeNB gateway control plane protocol stacks.

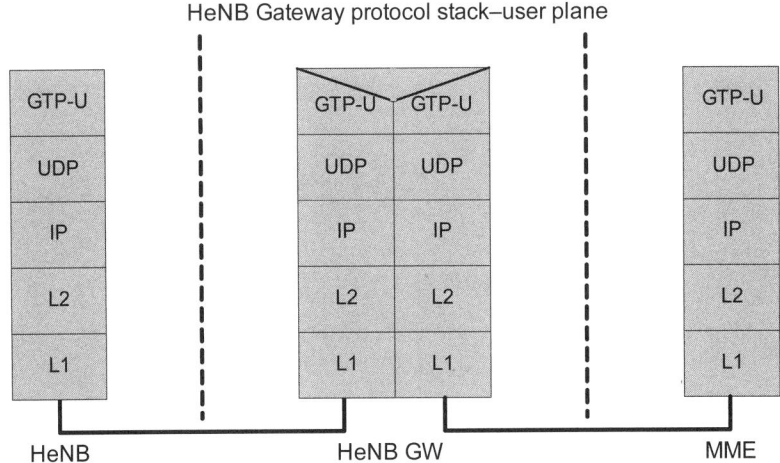

FIGURE 7.8
HeNB gateway user plane stacks.

mobile core network. This will result in less congestion in the EPC and a shorter round trip time (RTT) for user data. The main technologies standardized in 3GPP for traffic offloading are local IP access (LIPA) and selected IP traffic offload (SIPTO) [1, 13–15].

Note that from a mobile operator point of view, LIPA and SIPTO are useful technologies not only because they ease the network load but also because with LIPA and SIPTO (and other similar mechanisms yet to be standardized) the operators can keep the mobile user as their customer even when the UE is connected to a local network printer or browses the Internet. In principle, the mobile user could do both of those things without connecting to the mobile network, signaling-wise, but by adding LIPA and SIPTO into its service palette, the mobile operator can stay in control and can even charge for the usage of those services. This is very important for operators since users are increasingly accessing the services in the Internet with mobile devices that are not traditional mobile phones and may not have a subscription with a mobile operator. In this context, LIPA and SIPTO can be seen as technologies that can help mobile operators to keep mobile users as their customers.

7.4.5.1 LIPA

From Release 10 onward, HeNBs can support LIPA. However, UEs do not have to be Release 10 conformant in order to use LIPA; older mobiles can also enjoy this improvement if the network supports it. Using LIPA, an IP-capable UE that is connected to an HeNB can connect to other IP capable entities in the same local IP network via a local gateway (L-GW). Data traffic for LIPA does not traverse through the EPC (but the signaling does). The decision whether or not the HeNB operates in the LIPA mode remains under the control of the mobile network operator.

For a LIPA packet data network connection, the HeNB sets up and maintains an S5 connection to the EPC. This S5 interface does not go via the HeNB GW. The mobility of the LIPA PDN connection is not supported; the LIPA connection is always released at outgoing handover. The L-GW function in the HeNB triggers the release over the S5 interface. See Figure 7.9 for the LIPA architecture.

7.4.5.2 SIPTO

SIPTO stands for selected IP traffic offload. This term means that some IP traffic from HeNB can be routed directly to the Internet, therefore bypassing the EPC. One major difference between LIPA and SIPTO is that whereas LIPA can operate with HeNBs only, SIPTO can offload traffic from both HeNBs and eNodeBs.

For SIPTO there are three different architectures specified.

SIPTO Above RAN
This SIPTO function enables a mobile operator to offload certain types of traffic at a L-GW close to the (H)eNB that the UE is attached to. SIPTO above RAN architecture (Figure 7.10) selects a L-GW that is near the UE's point of attachment. The obvious problem with this approach is that it does not reduce any traffic in RAN since the L-GW is in the EPC. SIPTO above RAN was first specified in Release 10.

SIPTO at the Local Network with L-GW Collocated with the (H)eNB.
In this SIPTO architecture, the L-GW is collocated with the (H)eNB (see Figure 7.11). The MME shall use the LGW address proposed by the (H)eNB in the S1-AP message, instead of selecting the PGW via DNS interrogation. The S-GW remains in the mobile operator's core network. In this

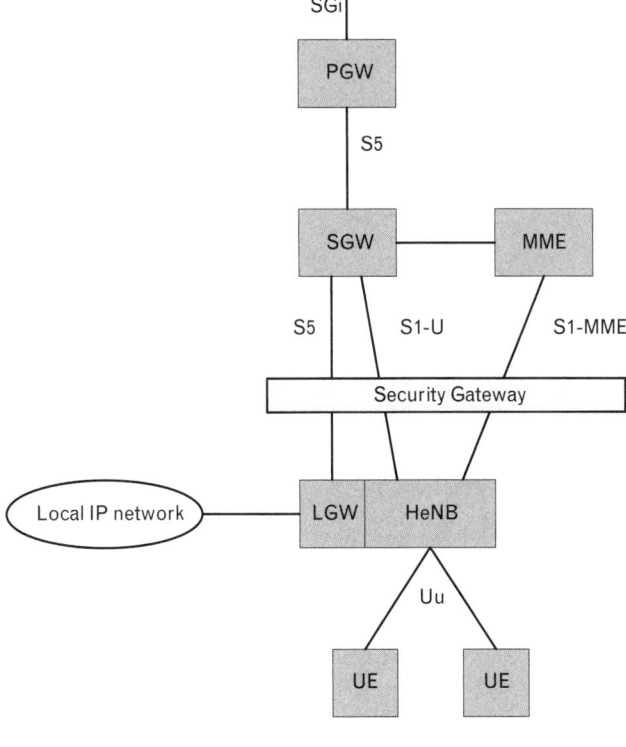

FIGURE 7.9
LIPA architecture.

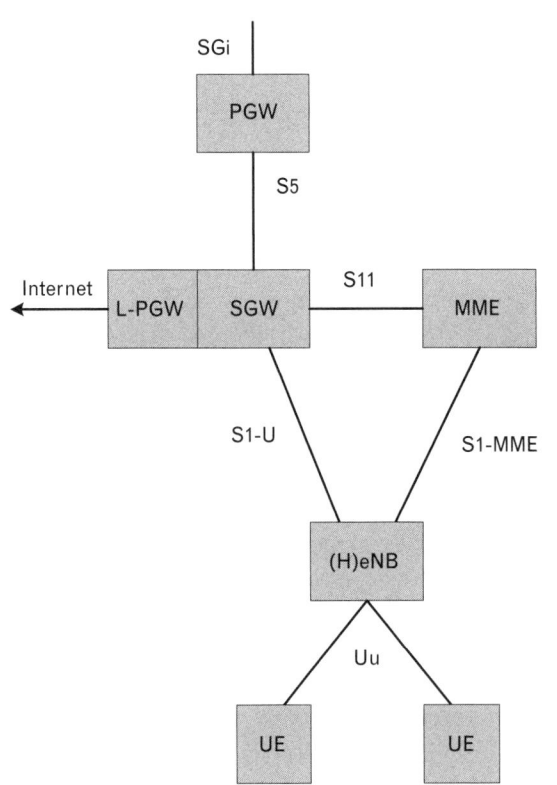

FIGURE 7.10
SIPTO architecture—SIPTO above RAN.

FIGURE 7.11
SIPTO architecture—SIPTO@LN with L-GW function collocated with the HeNB.

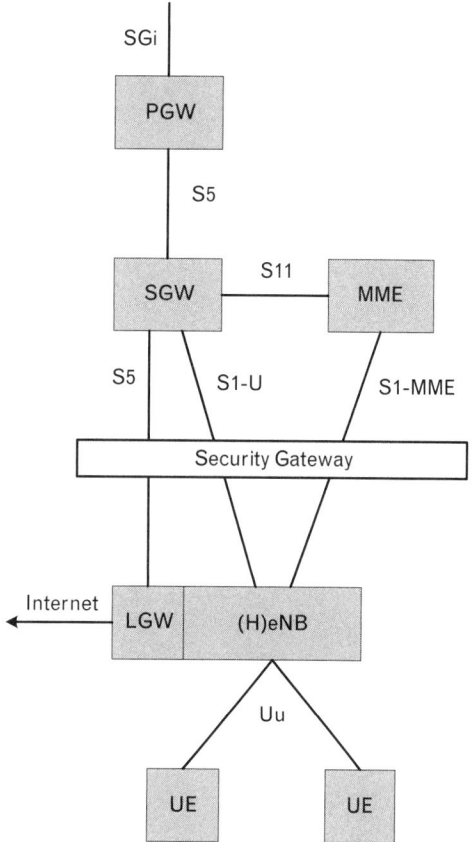

solution, the IP traffic is already sent to the Internet from the local (H)eNB, so it does not load the mobile network unnecessarily. This version of SIPTO was specified in Release 12.

SIPTO at the Local Network with Standalone GW (S-GW and L-GW Collocated)
In this SIPTO architecture, the stand-alone GW resides in the local network (see Figure 7.12). This GW includes both the L-GW and the S-GW functionality. The gateway and (H)eNBs can be separate nodes (i.e., one gateway can serve a number of (H)eNBs).

Also in this solution, the IP traffic is already sent to the Internet from RAN, though not directly from the local (H)eNB, so the IP traffic load in the mobile network is greatly reduced. This version of SIPTO was specified in Release 12.

FIGURE 7.12
*SIPTO architecture -
SIPTO@LN.*

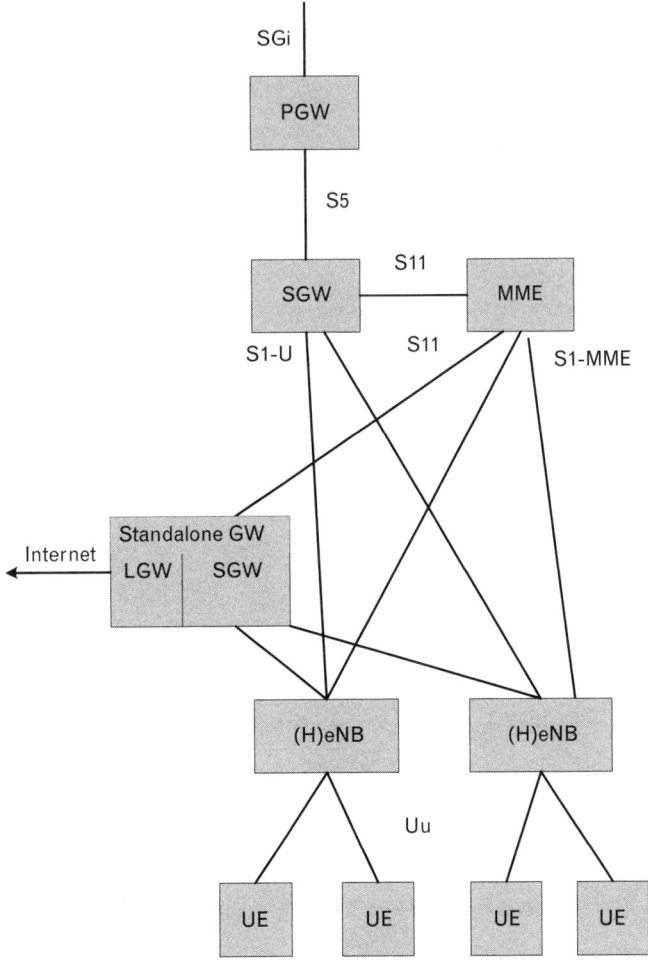

7.5 Relay Node

Relay node (RN) [2, 6] is an eNodeB that is wirelessly connected to an eNodeB serving the RN. RNs can be used to improve the coverage of the mobile network (e.g., in radio blackspots or areas where it is difficult to build conventional fiber backhaul).

The RN can be either inband or outband type. An inband RN uses the same carrier frequency for both the backhaul and access links. An outband RN employs different frequencies for both links. The inband RN will probably be the most common type in use.

This serving eNodeB is called the donor eNodeB (DeNB). The RN looks like a "real" eNodeB to a UE, and like a UE to the DeNB. The RN and the DeNB are connected by an air interface that is a modification of the standard LTE air interface Uu. We call this modified interface Un. Because the RN is an eNodeB, its backhaul toward the DeNB should also include

S1- and X2-interfaces. The network architecture for a RN, including the interfaces, is depicted in Figure 7.13.

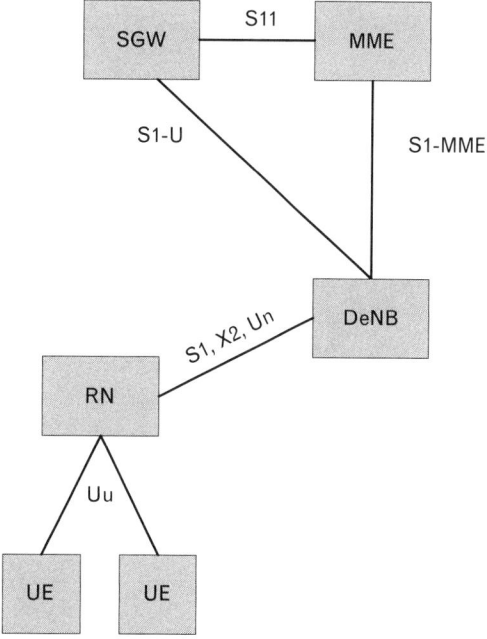

FIGURE 7.13
Relay node network architecture.

FIGURE 7.14
RN-DeNB S1 interface—control plane.

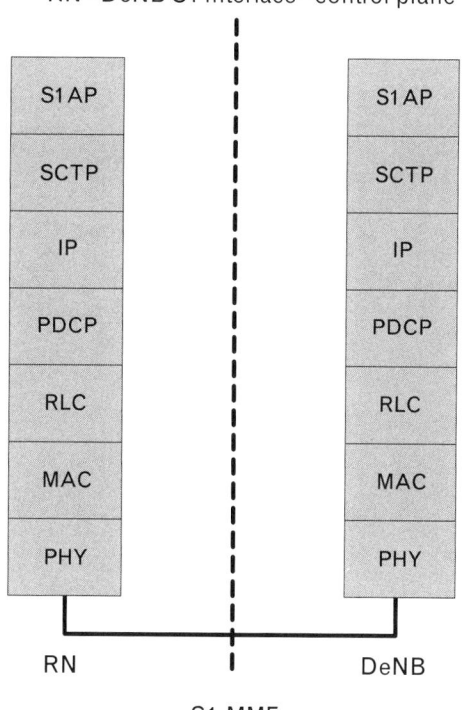

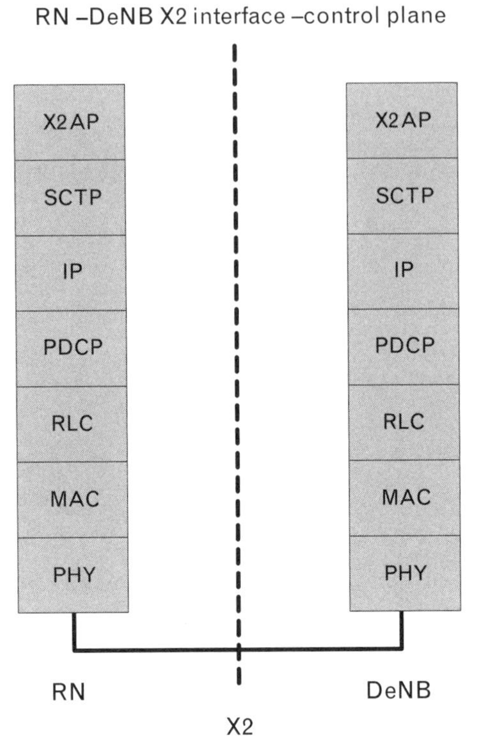

FIGURE 7.15
RN-DeNB X2 interface—control plane.

RN −DeNB X2 interface −control plane

X2 AP	X2 AP
SCTP	SCTP
IP	IP
PDCP	PDCP
RLC	RLC
MAC	MAC
PHY	PHY

RN — DeNB

X2

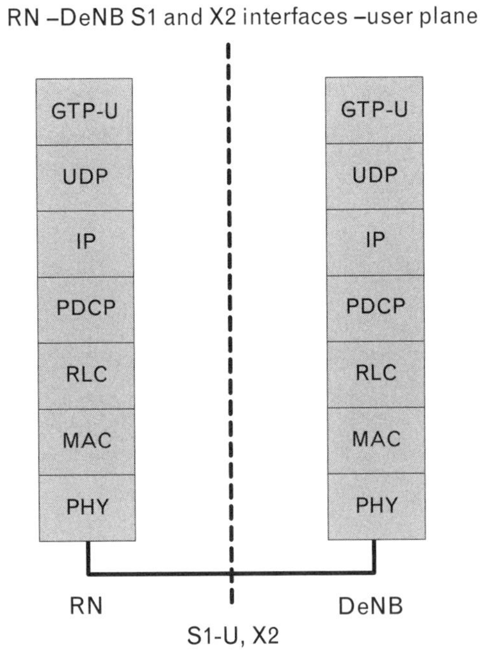

FIGURE 7.16
RN-DeNB S1 and X2 interfaces—user plane.

RN −DeNB S1 and X2 interfaces −user plane

GTP-U	GTP-U
UDP	UDP
IP	IP
PDCP	PDCP
RLC	RLC
MAC	MAC
PHY	PHY

RN — DeNB

S1-U, X2

Because the interface between the RN and the DeNB has to support several logical interfaces, the number of protocols involved is also large (see Figures 7.14, 7.15, and 7.16).

LTE-A has supported relays from Release 10 onward. Relay nodes are a new LTE-A feature, and they are discussed in depth in Chapter 11.

REFERENCES

[1] 3GPP TS 23.401, v. 12.2.0, General Packet Radio Service (GPRS) Enhancements for Evolved Universal Terrestrial Radio Access Network (E-UTRAN) Access; 09/2013.

[2] 3GPP TS 36.300, v 11.7.0, Evolved Universal Terrestrial Radio Access (E-UTRA) and Evolved Universal Terrestrial Radio Access Network (E-UTRAN); Overall Description; Stage 2; 09/2013.

[3] 3GPP TS 36.410, v. 11.1.0, Evolved Universal Terrestrial Radio Access Network (E-UTRAN); S1 General Aspects and Principles; 09/2013.

[4] 3GPP TS 36.420, v. 11.0.0, Evolved Universal Terrestrial Radio Access Network (E-UTRAN); X2 General Aspects and Principles; 09/2012.

[5] 3GPP TS 22.220, v.11.6.0, Service Requirements for Home Node B (HNB) and Home eNode B (HeNB); 09/2012.

[6] 3GPP TS 36.216, v. 11.0.0, Evolved Universal Terrestrial Radio Access (E-UTRA); Physical Layer for Relaying Operation; 09/2012.

[7] 3GPP TS 36.133, v. 12.1.0, Evolved Universal Terrestrial Radio Access (E-UTRA); Requirements for Support of Radio Resource Management; 09/2013.

[8] 3GPP TS 33.401, v 12.9.0, 3GPP System Architecture Evolution (SAE); Security Architecture; 09/2013.

[9] 3GPP TS 33.320, v 12.0.0, Security of Home Node B (HNB)/Home Evolved Node B (HeNB); 09/2013.

[10] IETF 5795: "The Robust Header Compression (ROHC) Framework"; 03/2010.

[11] 3GPP TS 36.331, v 11.5.0, Evolved Universal Terrestrial Radio Access (E-UTRA); Radio Resource Control (RRC); Protocol Specification; 09/2013.

[12] 3GPP TS 22.268, v 12.2.0, Public Warning System (PWS) Requirements; 06/2013.

[13] 3GPP TR 23.829, v 10.0.1, Local IP Access and Selected IP Traffic Offload (LIPA-SIPTO); 10/2011.

[14] 3GPP TR 23.859, v 12.0.1, Local IP Access (LIPA) Mobility and Selected IP Traffic Offload (SIPTO) at the Local Network; 04/2013

[15] NEC whitepaper, "MobileTraffic Offload—NEC's Cloud Centric Approach to Future Mobile Networks," NEC Corporation, www.nec.com/lte. Last accessed October 20, 2013.

CHAPTER 8

Core Network: Evolved Packet Core

8.1 Introduction

So far this book has discussed LTE and LTE-advanced. The latter term especially is becoming synonymous with the 4G mobile communications network. However, strictly speaking LTE (and LTE-A) cover only the access network evolution (i.e., E-UTRAN). A mobile communications system also needs a core network to function. The core network consists of equipment such as switches, gateways, and registers, and the transmission medium between those. In a similar way as the access network, also the 3GPP core network has undergone a major evolution so that it can support the packet-based access network and the new services it can provide. This evolution is known as the system architecture evolution (SAE).

The concepts and their names around core networks are often misused or misinterpreted. SAE and LTE should be used in equal terms: SAE is the core network evolution process and LTE is the access network evolution. However, quite often everything new in 3GPP systems is labeled as LTE, at least by people outside the 3GPP community. The LTE process has produced E-UTRAN. In a similar way the SAE process has produced the evolved packet core (EPC), the new packet-only core network system. Together E-UTRAN and EPC form the evolved packet system (EPS), which covers everything from UEs to registers and switches in the core network. Unfortunately, EPS as a concept is not well known, and not many people can connect that abbreviation to the latest mobile telecommunications system by 3GPP. Therefore, the name of this book is "Introduction to 4G Mobile Communications" and not "Introduction to Evolved Packet System".

It is also possible, at least in theory, to attach the EPC into a non-3GPP access network. Such networks could include, for example, CDMA2000, mobile WiMAX, and WLAN access networks. However, attaching a non-3GPP access network to the EPC is never a simple plug-and-play exercise. Reference [1] presents a long list of architecture enhancements and additional functions that are needed in the EPC in case a non-3GPP access network were to be employed.

The EPC functionality in case of non-3GPP access also depends on whether the non-3GPP access is regarded as trusted or untrusted (i.e., whether the core network operator believes that the access network can provide a high enough security level so that the UEs can access the EPC via this network without additional security at the EPC). If the network operator believes that the access network is untrusted, then the access from that network will take place through an additional network entity called the evolved packet data gateway (ePDG). The ePDG provides security mechanisms such as IPsec tunneling of UE connections over an untrusted access network.

8.2 Architecture

In this chapter we will introduce the EPC, its components and interfaces, its protocols, and its most important tasks. Figure 8.1 shows the overall EPC architecture. This diagram also depicts E-UTRAN, which is outside the EPC, but this is just to show how various entities in the EPC connect to E-UTRAN via different interfaces. The EPC is also often simply called the core network (CN). In fact, core network is a generic name, EPC is the SAE- (or LTE-) specific CN.

Note that the architecture diagram shows the functional entities in the EPC architecture. Whether these entities exist as separate physical entities is up to the network operator to decide. For example, the MME and the S-GW can be implemented as a single physical node; SGW and PDN GW is also a common combination. Depending on the services provided, several additional functional entities may be added to the diagram. For a full list of different entities in an EPC network, please consult [2].

The EPC architecture in Figure 8.1 consists of the following entities:

- Mobility management entity (MME);
- Serving gateway (S-GW);
- Packet data network gateway (P-GW);
- Home subscriber server (HSS);
- Evolved serving mobile location center (E-SMLC);
- Gateway mobile location center (GMLC);
- Policy control and charging rules function (PCRF).

These entities are briefly discussed next.

8.2.1 Mobility Management Entity

The MME is probably the most important single entity in the EPC because all C-plane signaling traffic between the UE and the EPC will go through it.

8.2 ARCHITECTURE

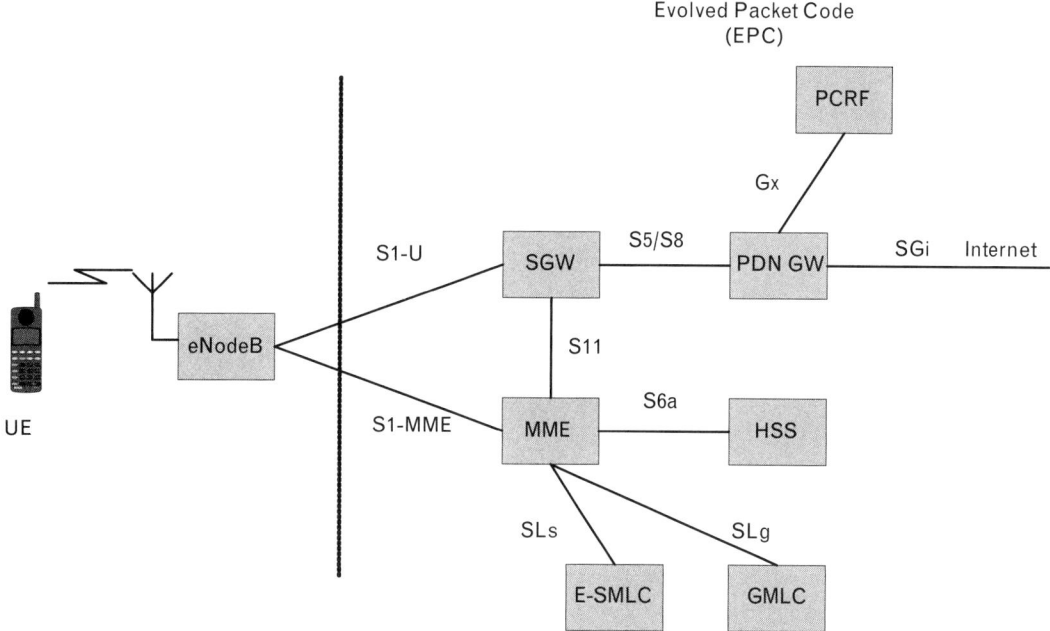

FIGURE 8.1 *EPC architecture.*

The MME controls, for the most part, what is happening in the E-UTRAN. The list of MME functions includes, among others:

- Tracking the UE in the idle mode;
- Paging the UE;
- UE authentication;
- Selection of the S-GW at the UE initial attach;
- Selection of the P-GW;
- Allocation of a wide variety of different temporary identities for the UE;
- Termination NAS signaling protocols;
- NAS signaling security;
- Support for the lawful interception (LI) of signaling traffic;
- Bearer management;
- Support for PWS message transmission;
- Selection of the SGSN in handovers to UMTS networks;
- MME selection for handovers that result in MME change;

- Roaming;
- Suppory for relaying.

And the list continues. Many of these functions actually require cooperation from several entities. Chapter 9, Procedures, gives several examples of how the different nodes in the network interact with each other in order to provide certain functionality.

8.2.2 Serving GW

Whereas the MME is the first contact point in the EPC for access network control plane signaling, the serving gateway (S-GW) fulfills the same function for user plane data. All user data to and from a UE will go through the S-GW; it is the first point of interconnect between the E-UTRAN and the EPC. The S-GW routes the incoming and outgoing IP packets to their correct destinations. Each UE is assigned a S-GW when the UE does the initial attach, and there cannot be more than one S-GW assigned for a UE at a given time. The S-GW is logically connected to the other EPC data gateway, the PDN GW.

The serving gateway implements the following functions:

- The local mobility anchor point for inter-eNodeB handover;
- Mobility anchoring for mobility to/from UMTS;
- E-UTRAN idle mode downlink packet buffering and initiation of network triggered service request procedure;
- Lawful interception of user data;
- Packet routing and forwarding;
- Transport level packet marking in the uplink and the downlink;
- Accounting for interoperator charging;
- Event reporting (e.g., change of RAT) to the PCRF.

8.2.3 Packet Data Network Gateway

The packet data network gateway (PDN GW, or P-GW) connects the EPC and the external IP networks, such as the Internet. The PDN GW routes user data packets to and from the PDNs (i.e, it handles the transport of IP data traffic between the UE and the external networks).

Often, both the P-GW and the S-GW functionality are combined into a single device by network vendors. However, whereas there can be only one S-GW per UE, it is possible to have more than one P-GW per UE if the UE is accessing several PDNs simultaneously.

The PDN GW performs the following functions:

- Per-user-based packet filtering (e.g., by deep packet inspection);
- Lawful interception;
- IP address allocation for UEs;
- Transport level packet marking in the uplink and downlink, based on the quality of service class identifier (QCI) of the associated bearer;
- UL and DL service level charging, gating, and rate enforcement;
- DL rate enforcement based on the APN aggregate maximum bit rate (APN-AMBR).

The APN aggregate maximum bit rate (APN-AMBR) is the maximum allowed total non-GBR throughput to specific APN. It is specified interdependently for the uplink and the downlink.

8.2.4 Home Subscriber Server

The home subscriber server (HSS) is a database that contains user-related information. It also provides some support functions for the following:

- Mobility management;
- Call and session setup;
- User authentication;
- Access authorization.

The authentication center (AuC), which generates authentication and security keys, is often integrated into the HSS.

Other EPC entities consult HSS whenever they need some information about the user. IMSI is the main reference key to any data that is stored in the HSS. Table 8.1 gives an example of the data held in the HSS [3].

8.2.5 Evolved Serving Mobile Location Center

The evolved serving mobile location center (E-SMLC) is a server that supports different location services for target UEs, including positioning of UEs and delivery of assistance data to UEs. Positioning has become an important application for UEs for many reasons. The mobile user may want to know where he is; therefore, the position data in itself is a service. Positioning may

TABLE 8.1 HSS Data

Field	Description
IMSI	IMSI is the main reference key.
MSISDN	The basic MSISDN of the UE (presence of MSISDN is optional).
IMEI/IMEISV	International mobile equipment identity—software version number.
MME identity	The identity of the MME currently serving this UE.
MME capabilities	Indicates the capabilities of the MME with respect to core functionality (e.g., regional access restrictions).
MS PS purged from EPS	Indicates that the EMM and ESM contexts of the UE are deleted from the MME.
ODB parameters	Indicates that the status of the operator determined barring.
Access restriction	Indicates the access restriction subscription information.
EPS subscribed charging characteristics	The charging characteristics for the UE (e.g., normal, prepaid, flat-rate, and/or hot billing subscription).
Trace reference	Identifies a record or a collection of records for a particular trace.
Trace type	Indicates the type of trace (e.g., HSS trace and/or MME/serving GW/PDN GW trace).
OMC Identity	Identifies the OMC that shall receive the trace record(s).
Subscribed-UE-AMBR	The maximum aggregated uplink and downlink MBRs to be shared across all non-GBR bearers according to the subscription of the user.
APN-OI replacement	Indicates the domain name to replace the APN OI when constructing the PDN GW FQDN upon which to perform a DNS resolution. This replacement applies for all the APNs in the subscriber's profile.
RFSP index	An index to specific RRM configuration in the E-UTRAN.
URRP-MME	UE reachability request parameter indicating that UE activity notification from MME has been requested by the HSS.
CSG subscription data	The CSG subscription data is a list of CSG IDs per PLMN and for each CSG ID optionally an associated expiration date which indicates the point in time when the subscription to the CSG ID expires; an absent expiration date indicates an unlimited subscription. For a CSG ID that can be used to access specific PDNs via local IP access, the CSG ID entry includes the corresponding APN(s).
VPLMN LIPA allowed	Specifies per PLMN whether the UE is allowed to use LIPA.
EPLMN list	Indicates the equivalent PLMN list for the UE's registered PLMN.
Subscribed periodic RAU/TAU timer	Indicates a subscribed periodic RAU/TAU timer value.
MPS CS priority	Indicates that the UE is subscribed to the eMLPP or 1x RTT priority service in the CS domain.
UE-SRVCC capability	Indicates whether or not the UE is UTRAN/GERAN SRVCC capable.
MPS EPS priority	Indicates that the UE is subscribed to MPS in the EPS domain.
Each subscription profile contains one or more PDN subscription contexts:	
Context identifier	Index of the PDN subscription context.
PDN address	Indicates subscribed IP address(es).
PDN type	Indicates the subscribed PDN type (IPv4, IPv6, IPv4v6).
APN-OI replacement	APN level APN-OI replacement that has same role as UE-level APN-OI replacement but with higher priority than UE-level APN-OI replacement. This is an optional parameter. When available, it shall be used to construct the PDN GW FQDN instead of UE-level APN-OI replacement.
Access point name (APN)	A label according to DNS naming conventions describing the access point to the packet data network (or a wildcard).

TABLE 8.1 (CONTINUED)

Field	Description
SIPTO permissions	Indicates whether the traffic associated with this APN is prohibited for SIPTO, allowed for SIPTO excluding SIPTO at the local network, allowed for SIPTO including SIPTO at the local network, or allowed for SIPTO at the local network only.
LIPA permissions	Indicates whether the PDN can be accessed via local IP access. Possible values are LIPA-prohibited, LIPA-only, and LIPA-conditional.
EPS subscribed QoS profile	The bearer-level QoS parameter values for that APN's default bearer (QCI and ARP).
Subscribed-APN-AMBR	The maximum aggregated uplink and downlink MBRs to be shared across all non-GBR bearers that are established for this APN.
EPS PDN subscribed charging characteristics	The charging characteristics of this PDN subscribed context for the UE (e.g., normal, prepaid, flat-rate, and/or hot billing subscription). The charging characteristics are associated with this APN.
VPLMN address allowed	Specifies per VPLMN whether for this APN the UE is allowed to use only the PDN GW in the domain of the HPLMN or additionally the PDN GW in the domain of the VPLMN.
PDN GW identity	The identity of the PDN GW used for this APN. The PDN GW identity may be either an FQDN or an IP address. The PDN GW identity refers to a specific PDN GW.
PDN GW allocation type	Indicates whether the PDN GW is statically allocated or dynamically selected by other nodes. A statically allocated PDN GW is not changed during PDN GW selection.
PLMN of PDN GW	Identifies the PLMN in which the dynamically selected PDN GW is located.
Homogenous support of IMS voice over PS sessions for MME	Indicates per UE and MME if "IMS voice over PS sessions" is homogeneously supported in all TAs in the serving MME, homogeneously not supported, or support is nonhomogeneous/unknown.
List of APN - PDN GW ID relations (for PDN subscription context with wildcard APN):	
APN - P-GW relation #n	The APN and the identity of the dynamically allocated PDN GW of a PDN connection that is authorized by the PDN subscription context with the wildcard APN. The PDN GW identity may be either an FQDN or an IP address. The PDN GW identity refers to a specific PDN GW.

also be used to enhance many other services. For example, if the mobile user wants to find a certain type of restaurant, the mobile operator may give him a selection of locations to choose from if the operator knows the UE's location. The governmental organizations may also be interested in the UE location. In United States, the FCC E-911 requirements, phase 2, stipulates that the network operator must be able to provide the location of emergency callers, anywhere between 50 meters and 300 meters of their actual position, depending on whether GPS handset or network-based E-911 technology is used. Even though FCC E-911 phase 2 is a US requirement only, similar requirements are likely to be adopted by many other countries.

The E-SMLC can employ both UE- and eNodeB-based positioning methods, or a combination of those methods to estimate the location of the UE. It may order the UE or the eNodeB (or several eNodeBs) to perform positioning measurements and report the results to the E-SMLC. The method(s) chosen depend on UE and eNodeB positioning capabilities and the positioning requirements (especially accuracy). Some of the positioning measurements may require assistance data to be provided to the UE by the E-SMLC. Note that some measurements may have already taken place as part of other procedures, such as handover preparation. The E-SMLC combines all the received results and determines a single location estimate for the target UE. Additional information like accuracy of the location estimate and velocity of the UE may also be provided.

The E-SMLC is connected to the MME via the SLs interface. LTE positioning protocol (LPP) is an end-to-end location protocol running between the UE and the E-SMLC. LPP acts as a method-agnostic supporting protocol for various UE positioning methods:

- Observed time difference of arrival (OTDOA);
- Global navigation satellite system (GNSS) and assisted GNSS (A-GNSS);
- Enhanced cell ID methods.

ODTOA is a downlink positioning method based on measured time differences observed by the UE from different eNodeBs. GNSS is a generic name for different satellite-based positioning methods such as GPS, Galileo, GLONASS, and Beidou. Assisted GNSS means that the base station provides additional satellite positioning data to UEs so that as a result, UE's own satellite positioning process is much quicker.

Positioning is a service that is certainly going to become much more important in the future. Positioning is needed by everybody, and the number of various location-based services is increasing. However, the widespread adoption of positioning technology has also brought about questions on its effect on privacy. From a pure technical point of view, positioning technologies enable the UE location to be monitored and logged continuously. Is the network operator allowed to do this? Who has the access to positioning data? E911 stipulates that the location of the UE should be made available in case of an emergency calls, but how about other governmental usage of location data? Can the police track the suspect 24/7? Can the positioning data be sold if the UE identifying information has been removed it? Is it OK to track the UE without the user knowing it (e.g., a company tracking the business phones of its employees or parents tracking the mobile phone of their child)? These are questions that need answers, preferably sooner than later. Positioning data is valuable, and it should have well-defined legal protection.

8.2.6 Gateway Mobile Location Center

The gateway mobile location center (GMLC) is the first node in the EPC that an external location application accesses. The GMLC checks the authorization for the location request, and, if the result is positive, the GMLC sends the positioning request to the MME. The MME forwards the request to the E-SMLC, which is responsible for the actual positioning process. There may be more than one GMLC in a PLMN.

8.2.7 Policy Control and Charging Rules Function

The policy and charging rules function (PCRF) acts as a policy decision point for policy and charging control of service data flows and IP bearer resources. The applicable policy and charging control is communicated to the policy control enforcement function (PCEF), which is located in the PDN GW. The PCRF checks the user's subscription profile, and then provides the QoS authorization that tells the PCEF how a data flow should be treated so that the resulting QoS conforms with the user's subscription. See [2, 4] for further information.

8.3 EPC Interfaces and Protocols

EPC has a large number of different interfaces between its entities, and each of these interfaces need protocols to relay communication. This section presents the most important interfaces in the EPC. Again, it must be noted that these logical interfaces do not always exist in deployed networks; a vendor may have implemented several functional entities in the same physical device, in which case there are no interfaces between them.

The functional entities and interfaces are different in control and user planes. For example, MME is a pure control plane entity, and S-GW exists in the user plane. Therefore, the presentation in this section is divided in two parts. First we discuss the control plane and then the user plane.

8.3.1 EPC Control Plane

EPC interfaces are wired interfaces, and therefore many of the protocols in them are not specified by 3GPP. Because 3GPP is specialized in drafting wireless standards, re-inventing the wheel was not sensible. Therefore, the lower layers (physical and transport network layers) in many EPC interfaces are implemented using non-3GPP protocols. In fact, 3GPP specifications exist for all of these protocol layers, but, for example, 3GPP TS 36.411 (S1 Layer 1) simply states that: "The support of any suitable layer 1 technique—like point-to-point or point-to-multipoint techniques—shall not be prevented." That is, 3GPP specifications do not really stipulate what kind of layer 1 an operator should use in its S1 interface.

The same applies to all interfaces in the EPC: the physical layer and the data link layer are left for the network operator to choose and implement. The IP layer can support IPv4 and/or IPv6.

8.3.1.1 S1-MME Interface

The S1-MME interface is located between the eNodeB and the MME. It handles all signaling traffic between the E-UTRAN and the EPC.

The protocols in this interface are as follows:

- The S1 application protocol (S1AP): This is the application layer control protocol between the eNodeB and the MME. It is defined in [5]. This specification is large and contains many important functions.

- Stream control transmission protocol (SCTP): This protocol guarantees delivery of signaling messages between MME and eNodeB (S1). SCTP is defined in [6] (see Figure 8.2).

8.3.1.2 UE-MME Interface

The UE-MME interface is a special case because actually it is a combination of two individual interfaces, Uu and S1-MME. It is presented here because the nonaccess stratum (NAS) protocol is specified to be used over

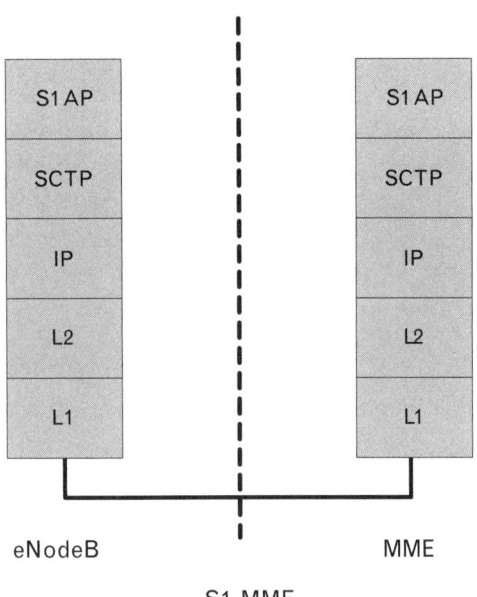

Figure 8.2
S1-MME interface.

the UE-MME interface. The NAS protocol is transparent to the eNodeB, which is situated between the Uu and S1-MME interfaces (see Figure 8.3).

The NAS protocol supports the following:

- Mobility management;
- U-plane bearer activation, modification, and deactivation;
- Ciphering and integrity protection of NAS signaling.

The NAS procedures are grouped in two categories:

- The EPS mobility management (EMM);
- The EPS session management (ESM).

The EPS mobility management protocol refers to procedures related to mobility over the E-UTRAN access, authentication, and security. The EPS session management protocol offers support to the establishment and handling of user data in the NAS. The NAS protocol is defined in [7].

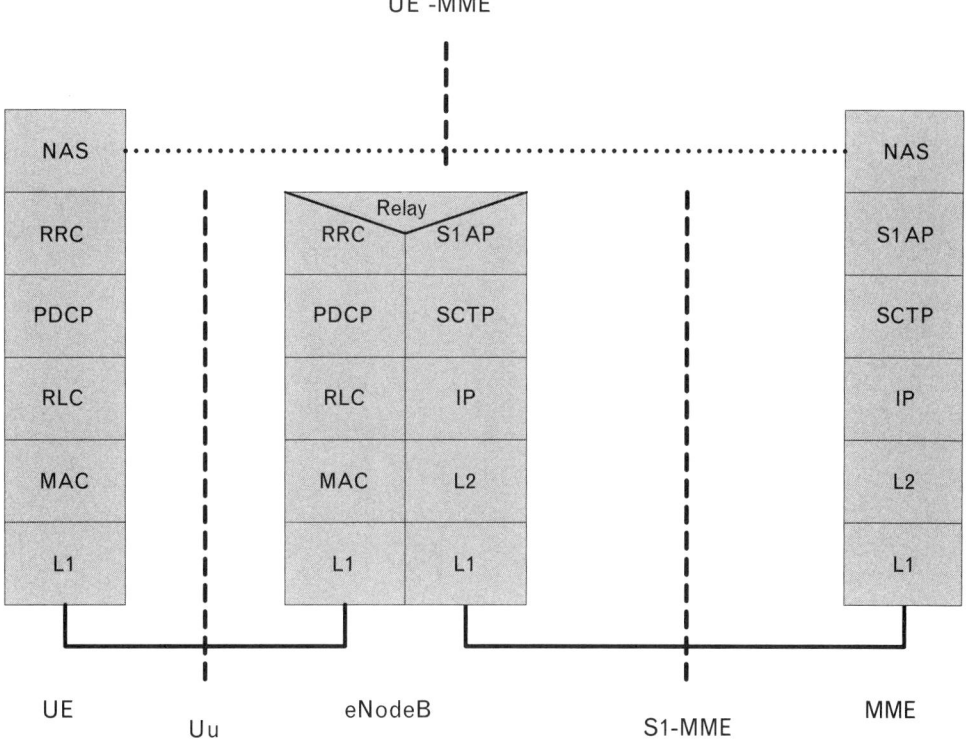

FIGURE 8.3 *UE-MME interface.*

8.3.1.3 S5/S8 Interface

This interface is located between the S-GW and the P-GW. It defines user plane tunneling and tunnel management between these entities. It is also used for the S-GW relocation due to UE mobility. The S8 interface is the inter-PLMN variant of the S5 interface (see Figure 8.4).

The protocols in this interface include the following:

- GPRS tunneling protocol for the control plane (GTP-C): This protocol tunnels signaling messages between S-GW and P-GW (S5 or S8). The GTP-C is defined in [8].
- User datagram protocol (UDP): This protocol transfers signaling messages between S-GW and P-GW. The UDP is defined in [9].

8.3.1.4 S10 Interface

This interface is located between two MMEs. It handles the MME relocation and MME to MME information transfer. Its protocols are the same as in the S5 interface (see Figure 8.5).

8.3.1.5 S11 Interface

This interface is located between the MME and the S-GW. Its protocols are the same as in the S5 interface (see Figure 8.6).

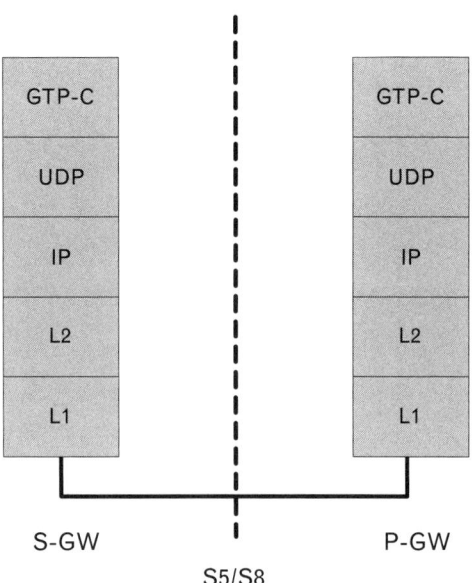

FIGURE 8.4
S5/S8 interface.

FIGURE 8.5
S10 interface.

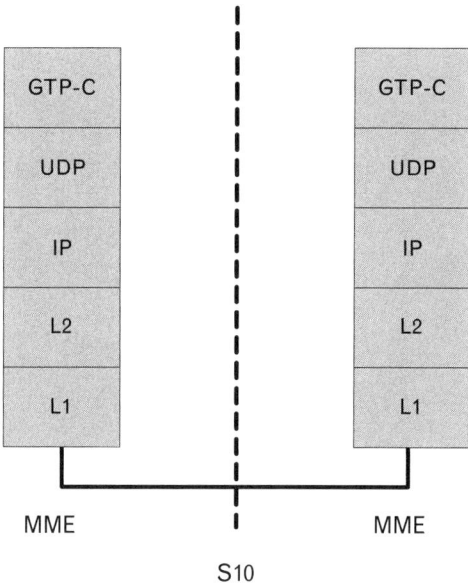

FIGURE 8.6
S11 interface.

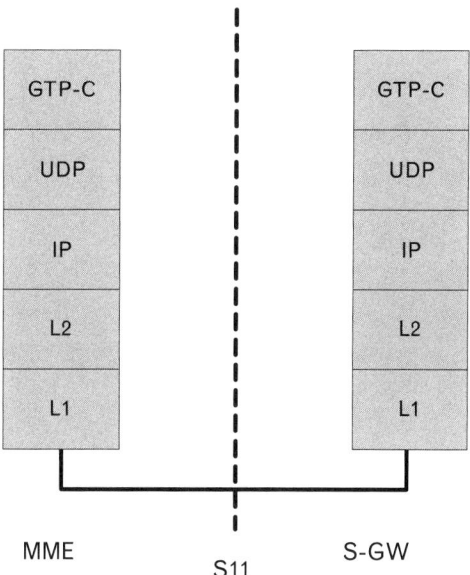

8.3.1.6 S6a Interface

This interface is located between the MME and the HSS. It is used to transfer subscription and authentication data for authenticating and authorizing user access (see Figure 8.7).

The protocols in this interface include the following:

FIGURE 8.7
S6a interface.

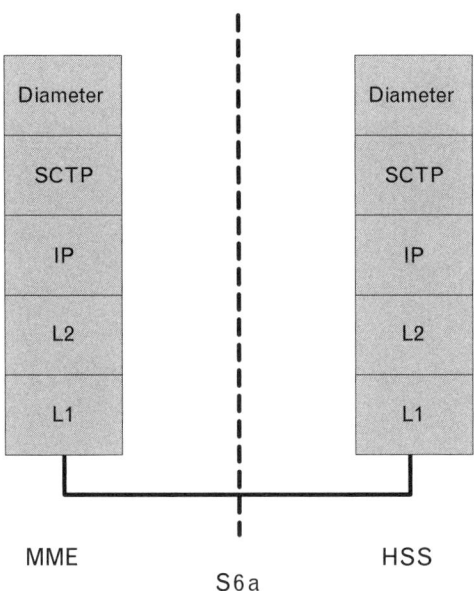

- Diameter: This protocol supports transferring of subscription and authentication data for authenticating and authorizing user access between the MME and the HSS. Diameter is defined in [10].

- Stream control transmission protocol (SCTP): This protocol transfers signaling messages. SCTP is defined in [6].

8.3.1.7 S13 Interface

This interface is located between the MME and the equipment identity register (EIR). It is used for UE identity checks (i.e., to check the identity of the equipment itself). The protocols in this interface are the same as in the S6a interface (see Figure 8.8).

8.3.1.8 S7a Interface

This interface is located between the MME and the CSG subscriber server (CSS). It carries CSG subscription data for roaming subscribers only. Its protocols are the same as in the S6a interface (see Figure 8.9).

8.3.2 EPC User Plane

As with control plane protocols, the lower layers in user plane protocols are left for the operator to choose. The IP layer can support IPv4 and/or IPv6.

FIGURE 8.8
S13 interface.

FIGURE 8.9
S7a interface.

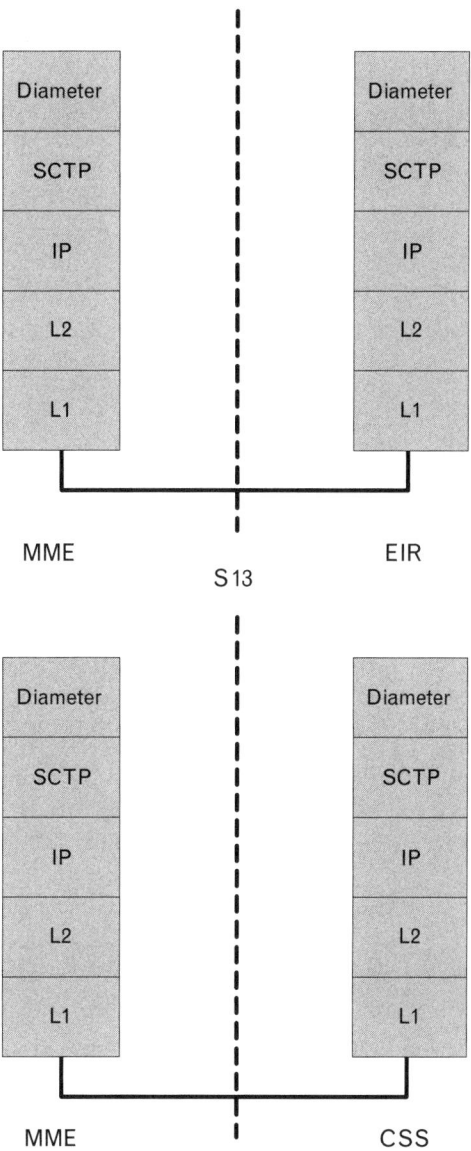

On top of the IP protocol layer, the user plane protocol stack employs the UDP protocol (since SCTP is for signaling only).

8.3.2.1 S1-U Interface

This interface is located between the eNodeB and the S-GW. It carries all user data between these two entities (see Figure 8.10).

FIGURE 8.10
S1-U interface.

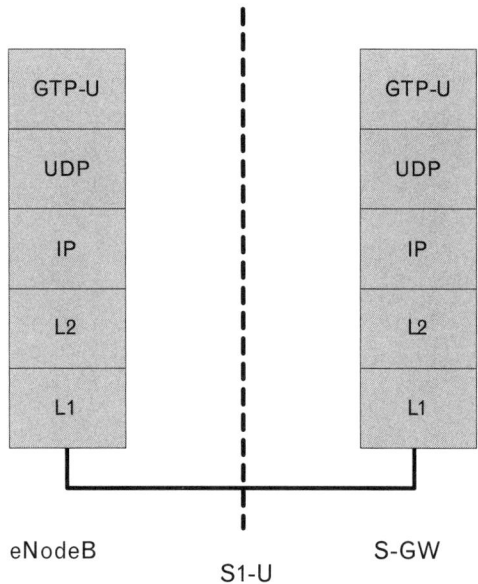

The protocols in this interface include GPRS tunneling protocol for the user plane (GTP-U). This protocol tunnels user data between eNodeB and the S-GW. The GTP-U encapsulates all end-user IP packets. It is defined in [11].

8.3.2.2 S4 Interface

This interface is located between the S-GW and the serving GPRS support node (SGSN) in the UMTS network. It provides control and mobility support between the GPRS core network and the S-GW. In addition, if the direct tunnel is not established, S4 provides user plane tunneling. S4 protocols are the same as in the S1-U interface (see Figure 8.11).

8.3.2.3 S12 Interface

This interface is located between the S-GW and the radio network controller (RNC) in UTRAN. It is used between UTRAN and the serving GW for user plane tunneling when the direct tunnel is established. The decision whether to use S12 is an operator configuration option. S12 protocols are the same as in the S1-U interface (see Figure 8.12).

8.3.3 Summary of EPC Interfaces

This section contains a list of other logical EPC interfaces and a short description of each. These, and other 3GPP-related interfaces, are discussed in [2].

8.3 EPC INTERFACES AND PROTOCOLS

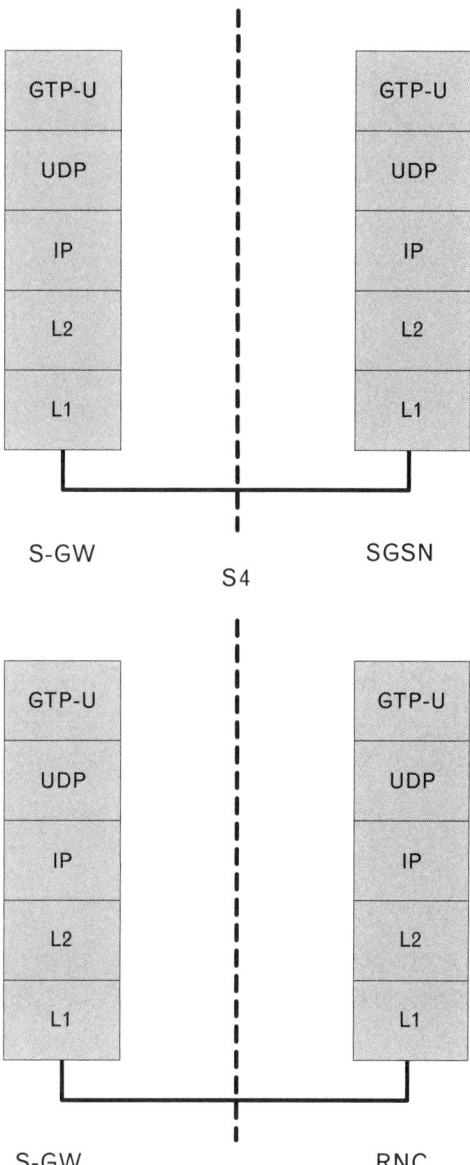

FIGURE 8.11
S4 interface.

FIGURE 8.12
S12 interface.

- S2a: the interface between trusted non-3GPP IP access and the PDN gateway. It provides the user plane with related control and mobility support. It is specified in [12] for proxy mobile IPv6 (PMIP) and [13] for client mobile IPv4 FA mode. When GTP is used in S2a, this interface is specified in [8] for GTP control plane (GTP-C) and in [14] for GTP user plane (GTP-U).

- S2b: the interface between the ePDG and the PDN gateway. It provides the user plane with related control and mobility support between ePDG and the gateway. This interface is specified in [8] and [14] for GTP and in [12] for PMIP.

- S2c: the interface between the UE and the PDN gateway. The protocol in the interface is based on the DS-MIPv6 protocol [15].

- S3: the interface between the MME and the SGSN. It enables user and bearer information exchange for inter-3GPP access network mobility in idle and/or active state. This interface is specified in [8].

- S6b: the interface between the PDN GW and the 3GPP AAA server/proxy. It is used for mobility-related authentication and authorization. It is specified in [16].

- S6c: the interface between the HSS and the SMS-GMSC/SMS-IWMSC/SMS router. It is used to interrogate the HSS of the required subscriber to obtain routing information for a short message directed to that subscriber. For further details, see [17, 18].

- S6d: the interface between the SGSN and the HSS. It is used to exchange the data related to the location of the UE and to the management of the subscriber. Signaling on this interface uses diameter S6a/S6d application as specified in [10].

- S7d: the interface between the CSS and the S4-SGSN. It is used to transfer the CSG subscription information stored in the VPLMN for roaming UEs. Signaling on this interface uses diameter S7d application as specified in [10].

- S9: the interface between the home PCRF and the visited PCRF. It handles the transfer of (QoS) policy and charging control information between the H-PCRF and the V-PCRF. This interface is specified in [19].

- S14: the interface between the UE and access network discovery and selection function (ANDSF). It is specified in [20].

- S101: the interface between the MME and the HRPD access network. Used for preregistration and handover signaling with the target system.

- S102: the interface between 3GPP2 1xCS IWS and the MME. Its usage is specified in [1, 21–23].

- S103: the interface between the S-GW and the HRPD HSGW. It is used to forward DL data during mobility from E-UTRAN to HRPD. It is specified in [1].

8.3 EPC INTERFACES AND PROTOCOLS

- Gx: the interface between the PCRF and the PCEF in the PDN GW. It handles the transfer of QoS policy and charging rules from PCRF to PCEF. This interface is specified in [24].

- Rx: the interface between the application function (AF) and the PCRF. It is used by the PCRF for the control of service data flows and IP bearer resources. For more information, see [4].

- SBc: the interface between the CBC and the MME for warning message delivery and control functions. This interface is specified in [25].

- SGd: the interface between the MME and the SMS-GMSC/SMS-IWMSC/SMS router. It is used to transfer short messages. This interface is specified in [17, 18]. Signaling on this interface uses diameter SGd application as specified in [18].

- SGi: the interface between the PDN GW and a packet data network. For more information, see [26].

- SWa: the interface between untrusted non-3GPP IP access and 3GPP AAA server/proxy. It transports access authentication, authorization, and charging-related information in a secure manner. It is specified in [16].

- STa: the interface between trusted non-3GPP IP access and 3GPP AAA server/proxy. It transports access authentication, authorization, mobility parameters, and charging-related information in a secure manner. It is specified in [16].

- SWd: the interface between the 3GPP AAA server and the 3GPP AAA proxy. It is specified in [16].

- SWm: the interface between the ePDG and 3GPP AAA server/proxy. It is used for AAA signaling. It is specified in [16].

- SWn: the interface between the ePDG and untrusted non-3GPP Access. It is specified in [16].

- SWu: the interface between the ePDG and the UE. It supports the handling of IPSec tunnels. It is specified in [20].

- SWx: the interface between the HSS and the 3GPP AAA server. It is used to transport authentication data. It is specified in [16].

- Uh: the interface between the UE and the CSG list server. It is specified in [27] for open mobil alliance (OMA) device management (DM) and in [28] for over the air (OTA).

References

[1] 3GPP TS 23.402, v 12.2.0, Architecture Enhancements for Non-3GPP Accesses; 09/2013.

[2] 3GPP TS 23.002, v 12.2.0, Network Architecture; 06/2013.

[3] 3GPP TS 23.401, v 12.2.0, General Packet Radio Service (GPRS) Enhancements for Evolved Universal Terrestrial Radio Access Network (E-UTRAN) Access; 09/2013.

[4] 3GPP TS 23.203, v 12.2.0, Policy and Charging Control Architecture; 09/2013.

[5] 3GPP TS 36.413, v 11.5.0, Evolved Universal Terrestrial Radio Access Network (E-UTRAN); S1 Application Protocol (S1AP); 09/2013.

[6] IETF RFC 4960, Stream Control Transmission Protocol, Stewart, R. (ed.), 09/2007.

[7] 3GPP TS 24.301, v 12.2.0, Non-Access-Stratum (NAS) Protocol for Evolved Packet System (EPS); Stage 3; 09/2013.

[8] 3GPP TS 29.274, v 12.2.0, 3GPP Evolved Packet System (EPS); Evolved General Packet Radio Service (GPRS) Tunneling Protocol for Control Plane (GTPv2-C); Stage 3; 09/2013.

[9] IETF RFC 768, User Datagram Protocol, Postel, J., 08/1980.

[10] 3GPP TS 29.272, v 12.2.0, Evolved Packet System (EPS); Mobility Management Entity (MME) and Serving GPRS Support Node (SGSN) Related Interfaces Based on Diameter Protocol; 09/2013.

[11] 3GPP TS 29.060, v 12.2.0, General Packet Radio Service (GPRS); GPRS Tunneling Protocol (GTP) Across the Gn and Gp Interface; 09/2013.

[12] 3GPP TS 29.275, v 12.0.0, Proxy Mobile IPv6 (PMIPv6) Based Mobility and Tunneling protocols; Stage 3; 09/2013.

[13] 3GPP TS 24.304, v 11.0.0, Mobility Management Based on Mobile IPv4; User Equipment (UE) - Foreign Agent Interface; Stage 3; 09/2012.

[14] 3GPP TS 29.281, v 11.6.0, General Packet Radio System (GPRS) Tunneling Protocol User Plane (GTPv1-U); 03/2013.

[15] 3GPP TS 24.303, v 11.3.0, Mobility Management Based on Dual-Stack Mobile IPv6; Stage 3; 06/2013.

[16] 3GPP TS 29.273, v 12.1.0, Evolved Packet System (EPS); 3GPP EPS AAA Interfaces; 09/2013.

[17] 3GPP TS 23.040, v 12.1.0, Technical Realization of the Short Message Service (SMS); 09/2013.

[18] 3GPP TS 29.338, v 12.2.0, Diameter Based Protocols to Support Short Message Service (SMS) Capable Mobile Management Entities (MMEs); 09/2013.

[19] 3GPP TS 29.215, v 12.1.0, Policy and Charging Control (PCC) over S9 Reference Point; Stage 3; 09/2013.

[20] 3GPP TS 24.302, v 12.2.0, Access to the 3GPP Evolved Packet Core (EPC) via Non-3GPP Access Networks; Stage 3; 09/2013.

[21] 3GPP TS 23.216, v 11.9.0, Single Radio Voice Call Continuity (SRVCC); Stage 2.

[22] 3GPP TS 23.272, v 12.0.0, Circuit Switched (CS) Fallback in Evolved Packet System (EPS); Stage 2; 09/2013.

[23] 3GPP2 A.S0008-C: "Interoperability Specification (IOS) for High Rate Packet Data (HRPD) Radio Access Network Interfaces with Session Control in the Access Network."

[24] 3GPP TS 29.212, v 12.2.0, Policy and Charging Control (PCC); Reference Points; 09/2013

[25] 3GPP TS 29.168, v 12.3.1, Cell Broadcast Center Interfaces with the Evolved Packet Core; Stage 3; 09/2013.

[26] 3GPP TS 29.061, v 12.3.0, Interworking Between the Public Land Mobile Network (PLMN) Supporting Packet Based Services and Packet Data Networks (PDN); 09/2013.

[27] 3GPP TS 24.285, v 11.0.0, Allowed Closed Subscriber Group (CSG) List; Management Object (MO); 09/2012.

[28] 3GPP TS 31.102, v 12.1.0, Characteristics of the Universal Subscriber Identity Module (USIM) Application; 09/2013.

CHAPTER 9

Procedures

9.1 Introduction

This chapter presents a selection of procedures that take place in E-UTRAN and EPC networks. The objective in this chapter has been to include procedures that show how the various entities in the system interact.

9.2 Cell Search

Cell search is the first procedure done by a UE when its power has been switched on. It is a procedure by which the UE acquires time and frequency synchronization with a cell and detects the cell ID of that cell. Without the initial synchronization, the UE cannot read system information broadcasts and thus cannot continue to perform any other procedures.

After the UE switch-on, the UE measures the wideband received power from several frequencies over a set of frequency bands. The frequencies are ranked based on their received signal strength indicator (RSSI), and the UE attempts the cell search starting from the best cell, using the downlink synchronization channels.

Note that in practical UE implementations, when the UE is switched off, it stores on the SIM card the cell search parameters of the last cell it was camped on. When the UE is again switched on, it may first try to see if the last cell is still available. If not, then the UE will go for the full cell scan as described in the previous paragraph. In most cases this arrangement works, noticeably cutting the time when the UE is out of service.

The eNodeB transmits two types of signals in order to help the UE with the synchronization process: the primary synchronization signal (PSS) and the secondary synchronization signal (SSS). The PSS and SSS are transmitted over the center 72 subcarriers in the first and sixth subframe of each frame. The PSS is sent in the last symbol of slots number 0 and 10, and the SSS as the next-to-last symbol in the same slots (see Figure 9.1).

There are three different PSS sequences, and each cell transmits only one of them. Once the UE detects the correct PSS sequence, it knows both the slot timing and the cell identity within a cell group (the three PSS

FIGURE 9.1
PSS and SSS location in the frame structure.

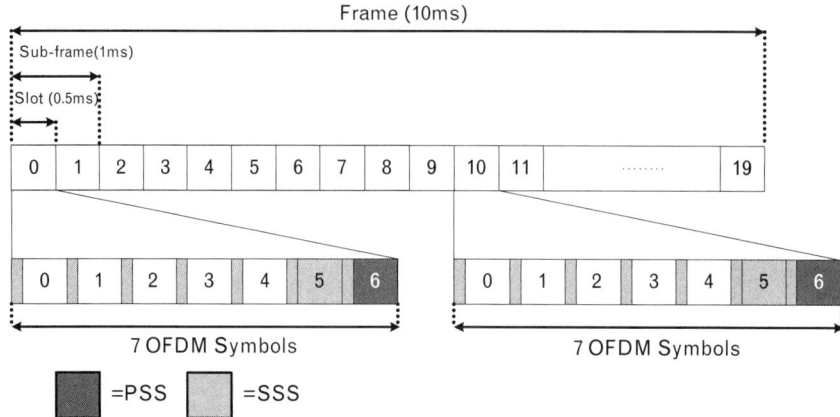

sequences map into three cell IDs). Thereafter, the UE cross-correlates the same channel with 168 possible SSS sequences. The contents of the PSS are the same in every slot it is transmitted, but SSS contents are alternated: different coding is used for SSS in slot number 0 and in slot number 10. However, the data contents of the SSS sequence are the same in both cases. The purpose of this arrangement is that once the UE manages to find the right SSS sequence, it knows at once where the frame boundary lies because it knows whether this slot is number 0 or number 10. Also, the separation in time of PSS and SSS symbols indicates whether the cell employs a normal or an extended cyclic prefix (CP) (see Figure 9.2).

As mentioned before, the SSS sequence also indicates the cell ID group number (0–167). By combining the cell group number, and the cell ID within the group (obtained from the PSS), the UE knows the physical cell identity (PCI). The PCI has a very important property in that it indicates the location of cell-specific reference signals in downlink frames. These reference signals are used for channel estimation.

FIGURE 9.2
CP length detection with PSS and SSS.

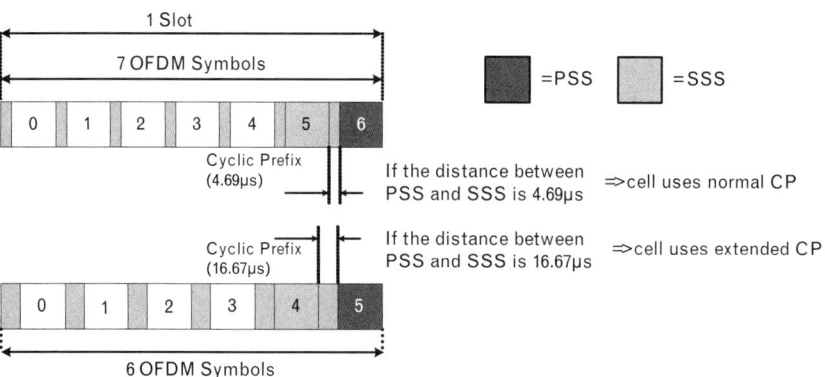

There are all together 504 locally unique PCIs. The network operator should take care that the same PCI is not allocated for two cells that are located near each other, because this may lead to PCI confusion: the UE cannot distinguish the downlink transmissions of two cells with the same PCI. Normally in a macro-cell network the number of PCIs is large enough that the PCI confusion can be avoided, but femtocell deployments may easily result in PCI confusion. This is especially true in the case of HeNBs because those devices can be deployed anywhere, and the HeNB owner cannot be assumed to know about technical details such as PCI confusion.

PSS and SSS sequences are sent only on the central 72 subcarriers of any channel, regardless of the total channel bandwidth. This arrangement allows the UE to perform the synchronization without any preknowledge of the channel bandwidth, because synchronization signals are always transmitted on same frequency positions in relation to the channel central frequency.

Once the UE has obtained time and frequency synchronization, and it knows the PCI, it can start receiving system information. The first system information message to be received is the master information block (MIB), which contains three important pieces of information:

- The downlink channel bandwidth;
- The system frame number;
- The physical hybrid ARQ indicator channel configuration info.

With this information the UE can receive and decode system information block type 1 and thereafter all other required system information blocks.

The description of the cell search procedure is divided between several specifications. The best place to start is [1].

9.3 Random Access

The random access procedure is performed when the UE does not yet have any connection with the network. With random access, the UE gains uplink time synchronization, which is prerequisite for any two-way air interface communication. Note that the UE can gain downlink time synchronization simply by listening the eNodeB and receiving synchronization signals, as described in Section 9.2.

A UE does not have uplink time synchronization with a cell, for example, in following cases:

- This is UE's initial access for this eNodeB.

- Existing connection has failed, and an RRC connection reestablishment is attempted.
- Handover to a new cell.

The random access can be either contention-, or noncontention-based. In noncontention-based random access, the UE has to use a random access preamble that is assigned to it by the eNodeB via dedicated signaling. Obviously this is possible only in special cases where there already exists a dedicated connection with the UE (e.g., in handovers when the preamble can be sent via the old cell).

Figure 9.3 depicts the contention-based random access procedure. In LTE, random access is a MAC layer procedure and therefore specified in [2]. Additional random access information can also be found from [1, 3–5].

1. The random access procedure starts when the UE sends a random access preamble on the PRACH channel. The preamble is actually a 6-bit-long random ID. Each cell has 64 different random access preambles, although not all of them are available for contention-based random access. The UE selects one of them and transmits it in the next subframe that is allocated for the PRACH channel.
2. The eNodeB responds with a random access response (RAR) on the PDSCH using the same preamble it just received. The UE listens to this channel during the RAR time window, which starts at the subframe that contains the end of the preamble transmission plus three subframes and has a configurable length in subframes (see Figure 9.4). The response message also contains uplink timing

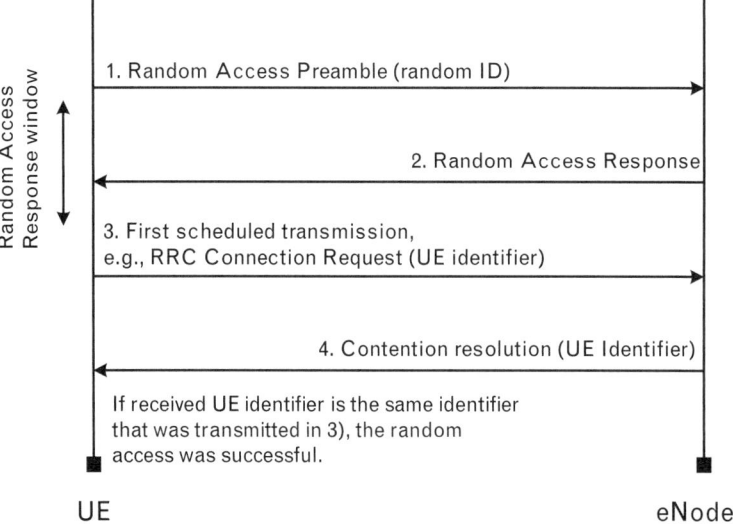

FIGURE 9.3
Random access, contention based.

9.3 RANDOM ACCESS

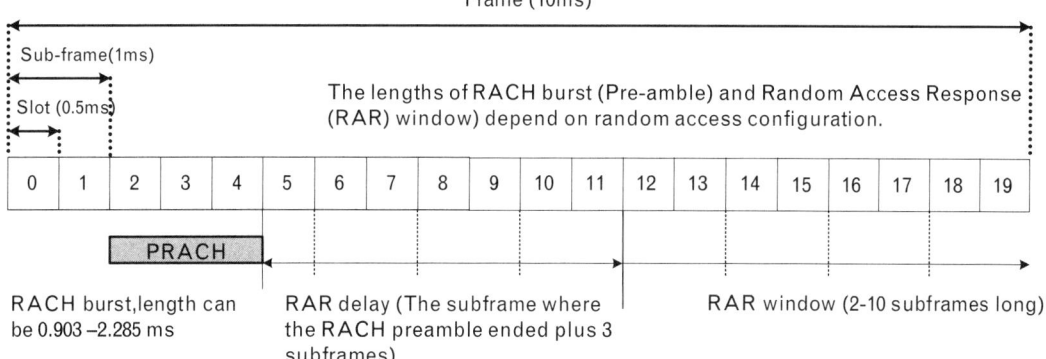

FIGURE 9.4 *Random access response timing.*

alignment information, uplink resource grant for the next uplink message, and a temporary cell radio network temporary identifier (C-RNTI). Note that it is possible that more than one UE got the same response message in case they has selected the same preamble in phase 1 and then transmitted it at the same time and at the same frequency. This clash can be resolved later, as shown in phase 4.

If the UE did not receive a response within the RAR time window, then it has to try again. A new random access preamble must be chosen, and the UE must delay the new preamble transmission by the parameter backoff time.

3. In phase 3 the UE sends its first scheduled uplink transmission on the PUSCH channel. This message contains the actual message for which the whole random access procedure was started (e.g., an RRC connection request). This message must also contain the UE NAS identifier if this procedure is an initial UE access or a C-RNTI if that was assigned to the UE earlier via dedicated signaling. (Note: therefore this C-RNTI is not the same identifier as the temporary C-RNTI assigned in phase 2.) It is important that a unique UE identifier is transmitted in this phase because the same identifier will be used for contention resolution in phase 4.

4. Contention resolution: the eNodeB sends a HARQ feedback message on the PDCCH using the UE identifier that it just received in phase 3. Since this identifier is unique and known to only one UE, other UEs will understand that a random access clash has happened and will stop their random access procedures.

9.4 Tracking Area Update

Tracking areas (TAs) are used in LTE to locate UEs for paging. The corresponding concepts in UMTS are location areas and routing areas. In UMTS both are needed because location area is a CS domain concept and routing area a PS domain concept. In LTE there is only one domain, packet switched, and so only tracking areas are needed.

A tracking area contains a group of cells. How big this group is and which cells belong to the same group depend on network planning. A group of TAs forms a tracking area identity (TAI) list. A TA is a parameter that is common to all users. It is broadcast in cell's system information messages. A TAI list, on the other hand, is a tailor-made parameter for each user. The reason for this arrangement is that a TA alone would be a poor method to keep track of a UE. It would be impossible to design TAs that would suit for all users because users' movement patterns are very different. If you design a TA for user A along her daily movement patterns, trying to minimize her TA boundary crossings, the same TA map may cause numerous TA updates for user B with his different movement patterns.

The TAI list contains a number of TA identities, up to 16 in Release 12. Instead of the network paging the UE in one TA, the UE is now paged in all TAs in the TAI list. The TAI list of each user is stored in the serving MME. The TAI list concept gives the network planner great flexibility. Now each user can be given her own tracking area scheme. The UE triggers a tracking area update only when it camps on a cell whose TA is not found from the UE's TAI list. The triggered tracking area update request contains the new TA to be added to UE's TAI list. It is up to MME to decide how many TA entries there are in UE's TAI list and according to which policy any older TAs are removed from it.

The smaller the TAI list is, the fewer paging messages are needed. Because a paging message has to be sent in all cells in all TAs in the TAI list, the number of paging messages could be very large with a large TAI list. This would suggest that a very small TAI list is preferable. If the tracking area covers only one cell, and the TAI list contains only this TA, then the paging has to be carried out only in one cell. However, the UE also has to perform a tracking area update every time it crosses a cell boundary to a TA that is not in its TAI list. With small TAI lists, this happens more often. Therefore, a network planner must find a good balance between suitably sized tracking areas and individually tailored TAI lists according to each user's movement and usage patters. This is a balancing act between the number of paging and tracking area update procedures. Moreover, selecting correct cells into a TA is important. For example, it is usually a good idea to group the cells along a busy road into the same tracking area. The TA and TAI list concept is depicted in Figure 9.5.

9.4 TRACKING AREA UPDATE 175

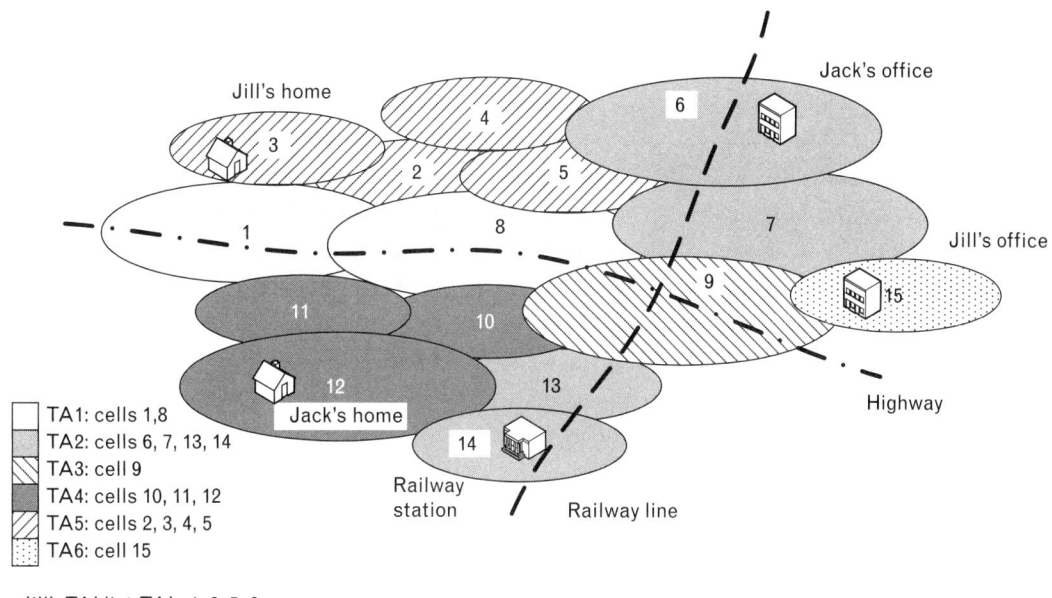

FIGURE 9.5 *Tracking areas and tracking area identity lists.*

In this example, Jill is commuting to her office daily using a route via cells 3, 1, 8, 9, and 15. Jack prefers trains for commuting and uses a route via cells 12, 14, 13, 9, 7, and 6. If Jill's TAI list contains TAs 1, 3, 5, and 6, her UE does not have to make any tracking area updates during her daily commute. Similarly Jack's TAI list 2, 3, and 4 gives him journeys without tracking area updates. However, the drawback is that a paging message has to be sent to all of these TAs. In Jill's cases this includes cells 1, 2, 3, 4, 5, 8, 9, and 15. A paging message to Jack has to be sent to cells 6, 7, 9, 10, 11, 12, 13, and 14.

A TA update can also be periodic. The network sends a value for timer T3412 via dedicated signaling, and, once this timer expires, the UE has to start a tracking area update. Note that this timer is UE specific, not cell or TA specific. Therefore, the network can adjust the value of this timer according to user's behavior. For example, there is no need to request frequent tracking area updates from a user who always stays within a one tracking area.

The example procedures here include two tracking area updates: first a basic update (Figure 9.6) and then a full tracking area update with MME and S-GW changes (Figure 9.7).

1. There are several reasons a tracking area update procedure can be triggered. However, the most usual cases are that in the idle mode the UE has camped on a new cell belonging to a tracking area that is not part of the UE's registered TAI list or that the periodic

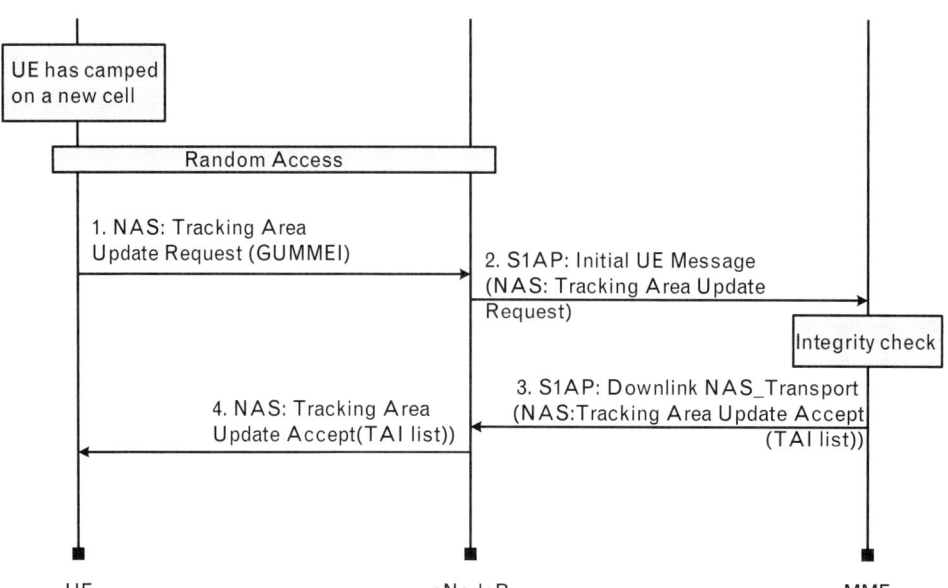

FIGURE 9.6 *Tracking area update—the basic case.*

TA update timer has expired. For other TAU triggering reasons, consult Section 5.5.3.1 of [6]. The procedure starts with the UE sending a NAS: Tracking_Area_Update_Request to the eNodeB. This message contains the current GUMMEI parameter. Before this message can be sent over the air interface, a connection must be established using the random access procedure.

2. The eNodeB checks the destination MME from the GUMMEI parameter. If that parameter is not available, or the destination MME is not associated with the eNodeB, the eNodeB has to select a new MME. The NAS: Tracking_Area_Update_Request is then sent over to the new MME within the S1AP: Initial_UE_Message. However, in this case there will not be a change of MME, so the message goes to the serving MME.

3. The MME performs an integrity check to the received tracking area update request message. If it fails, an authentication procedure will follow. Otherwise, the MME considers the procedure successful and sends a NAS: Tracking Area_Update_Accept message to the eNodeB, encapsulated into a S1AP: Downlink_NAS_Transport message. This message contains the new TAI list. Note that it is up to the MME implementation to decide how big the TAI list is and which TAIs it will include. The maximum size of the TAI list is 16 TAIs in release 12 specifications.

4. The eNodeB delivers the NAS: Tracking_Area_Update_Accept message further to the UE.

9.4 TRACKING AREA UPDATE

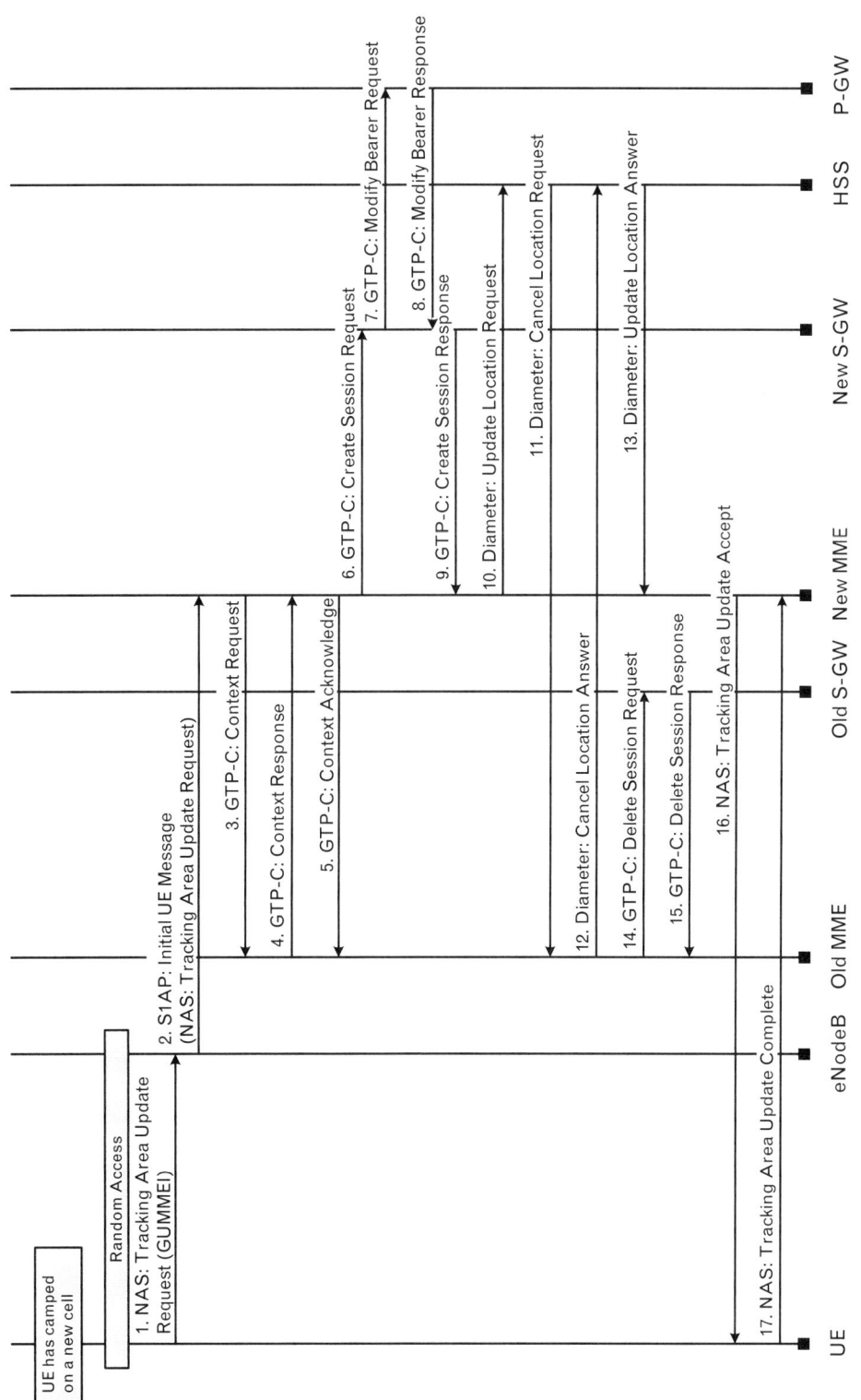

FIGURE 9.7 *Tracking area update with MME and S-GW changes.*

As seen, the tracking area update can be a simple process if both MME and S-GW remain the same in the procedure. This is the case in most tracking area update procedures if the S1-flex scheme is employed in the S1 interface. In S1-flex, several MMEs and S-GWs can form a pool in which they share the load within the pool members. One eNodeB is simultaneously connected to all MMEs and S-GWs within the pool. However, as long as the UE stays within the pool area, its context will normally remain with the same MME. This is to avoid unnecessary update procedures within the network every time the serving MME changes.

Tracking area update (TAU) with MME and S-GW changes is a much more complex procedure than the previous basic procedure. If S1-flex is employed, then this procedure will take place when the UE crosses the MME pool area edge.

1. Step 1 here is similar to step 1 in the previous example.
2. The eNodeB checks the destination MME from the GUMMEI parameter. If that parameter is not available or the destination MME is not associated with the eNodeB, the eNodeB has to select a new MME. Since this scenario is about a TAU with MME and S-GW changes, then the MME indicated by GUMMEI is not associated with the eNodeB, and a new MME must be selected. The eNodeB will send the NAS:Tracking_Area_Update_Request message within the S1AP: Initial_UE_Message to the new MME.
3. The MME uses the old GUMMEI to find the identity of the old MME and sends a GTP-C: Context_Request message to the old MME in order to retrieve the user information. The old MME starts a timer, which will be later used in step 14.
4. The old MME responds with a GTP-C: Context_Response message, which includes the stored UE context.
5. If the MME decides to change the S-GW, it will then send a GTP-C: Context_Acknowledge message to the old MME, indicating that the S-GW will change, too. On receiving this message, the old MME marks into its UE context that the related information in the S-GW and in the HSS is invalid.
6. The new MME compares the EPS bearer status received from the UE with the bearer contexts received from the old MME and releases any network resources related to EPS bearers that are not active in the UE. The new MME will then send a GTP-C: Create_Session_Request message to the new S-GW.
7. The S-GW informs the P-GW about the change with a GTP-C: Modify_Bearer_Request message.
8. The P-GW updates its bearer contexts and returns a GTP-C: Modify_Bearer_Response message.

9. The S-GW updates its own bearer context. This enables the SGW to route bearer PDUs to the P-GW when received from the eNodeB. After this the S-GW sends a GTP-C: Create_Session_Response message to the new MME.
10. The new MME checks whether it holds subscription data for the UE. If the answer is no, the new MME sends a Diameter: Update_Location_Request message to the HSS.
11. The HSS sends a Diameter: Cancel_Location_Request message to the old MME with the cancellation type set to "Update Procedure."
12. The old MME removes the MM context of this UE from its records. The old MME acknowledges the message with a Diamater: Cancel_Location_Answer message.
13. The HSS sends a Diameter: Update_Location_Answer message to the new MME, with the IMSI and UE subscription data. The new MME will then construct a context for the UE.
14. Once the timer that was started in step 3 expires, the old MME releases any local MME bearer resources and sends a GTP-C: Delete Session Request message to the old S-GW so that it can also delete all bearer resources related to the UE. The S-GW also discards any data packets buffered for the UE.
15. The S-GW acknowledges with a GTP-C: Delete_Session_Response message to the old MME.
16. The new MME sends a NAS:Tracking_Area_Update_Accept message to the UE. The GUMMEI is included if the MME allocated a new GUMMEI for the UE.
17. If the GUMMEI was included in step 16, then the UE acknowledges the received message with a NAS: Tracking_Area_Update_Complete message to the MME.

Tracking area updates are discussed in [7–9].

9.5 Initial Context Setup

Initial context setup is a procedure whereby the UE initiates a data connection (though the original request for that might have come from the network in the form of paging), and the network side then allocates necessary resources in different nodes, updates registers, and creates bearers/sessions/contexts so that the routing of messages for the new connection is clear for all participating nodes (see Figure 9.8).

1. The UE initiates the attach procedure by sending the NAS:Attach_Request message to the eNodeB. In the RRC layer this message

180 PROCEDURES

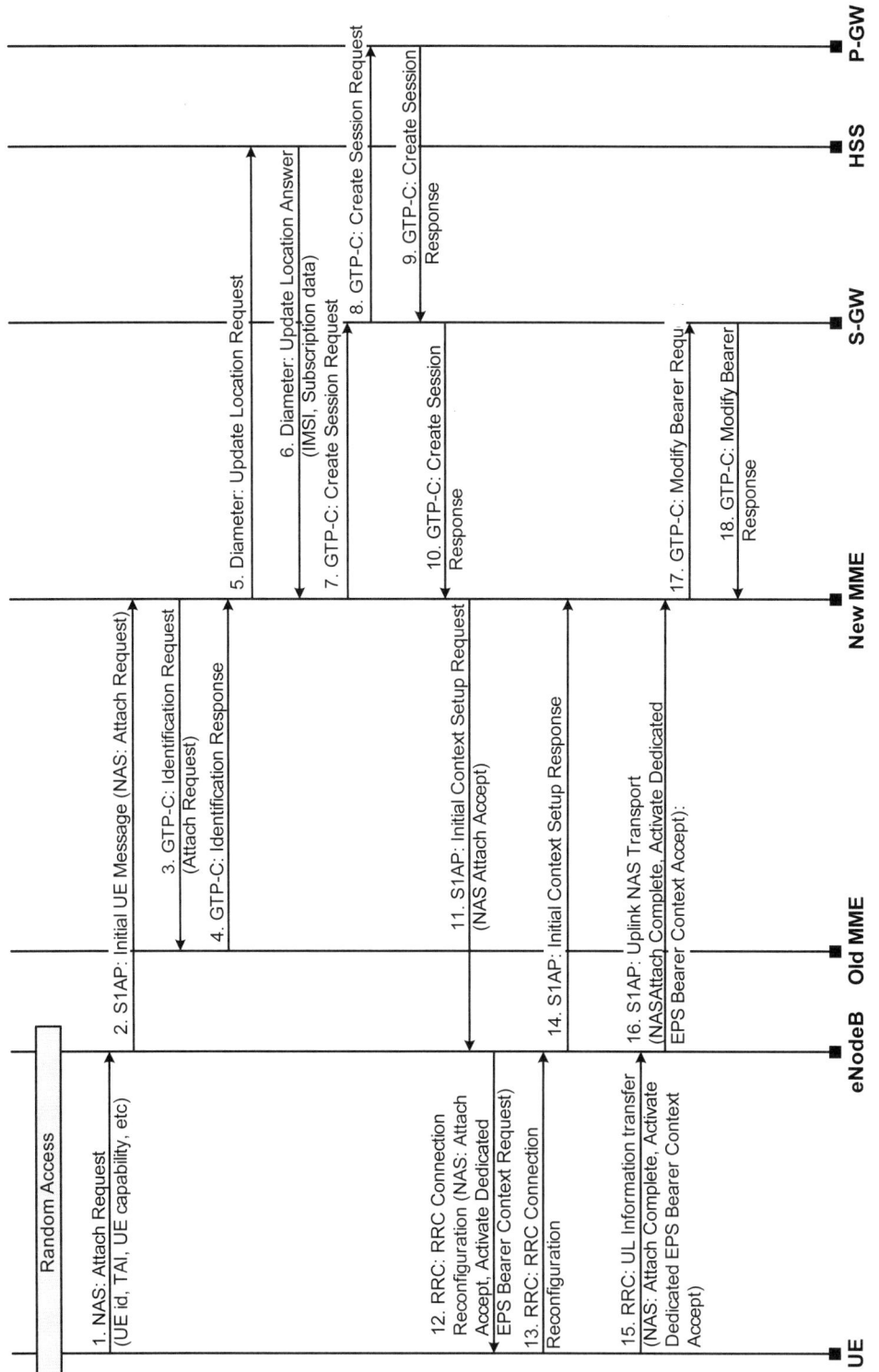

FIGURE 9.8 *Initial context setup.*

9.5 INITIAL CONTEXT SETUP

is encapsulated into the RRC: UL_Information_Transfer message. This message contains the UE identity, tracking area identity, UE capability, various security parameters, and so on. Before it can be sent over the air interface, the UE must establish a dedicated connection with the random access procedure.

2. The eNodeB checks the received message to find out if the UE is already associated with an MME (i.e., if there a GUMMEI parameter attached). Otherwise, the eNodeB selects an MME by using the MME selection function and then sends the S1AP: Initial_UE_Message to the MME.

3. If the GUMMEI in the received message does not match MME's own identity, then the MME has to send a GTP-C: Identification_Request message to the old MME (identified by the GUMMEI). This message includes the attach request.

4. The old MME verifies the attach request and sends back a GTP-C: Identification_Response message.

5. The new MME sends a Diameter: Update_Location_Request message to the HSS.

6. The HSS sends back a Diameter: Update_Location_Answer message that includes IMSI and subscription data.

   The new MME checks whether the UE is allowed to access the current tracking area and whether it has a valid subscription. If both cases are true, the MME creates a new context for the UE. The MME also selects a S-GW for the UE by using a S-GW selection function and allocates an EPS bearer identity. If the HSS did not indicate a P-GW address to the MME, then the MME also selects that address using a special selection function.

7. The MME sends a GTP-C: Create_Session_Request message to the S-GW, including the EPS bearer identity and the P-GW address (among other parameters).

8. The S-GW adds the new EPS bearer identity into the EPS bearer table and sends a GTP-C: Create_Session_Request message to the P-GW.

9. The P-GW adds a new entry into its EPS bearer table. It also generates a charging ID. After that the PDN-GW sends a GTP-C: Create_Session_Response message back to the S-GW.

10. The S-GW sends a GTP-C: Create Session Response message to the MME. Both the S-GW and the P-GW are now aware of the new bearer/session.

11. The MME sends an S1AP: Initial_Context_Setup_Request message to the eNodeB in order to set up the UE context. This message also contains the NAS: Attach_Accept message.

12. When the eNodeB receives the message, it creates a new radio bearer based on the information in the message and then sends a RRC: RRC_Connection_Reconfiguration message to the UE, including the new EPS radio bearer identity and the encapsulated NAS: Attach_Accept and NAS: Activate_Dedicated_EPS_Bearer_Context_Request messages.
13. The UE sends an RRC: RRC_Connection_Reconfiguration_Complete message to the eNodeB to acknowledge the successful establishment of the radio bearer.
14. The eNodeB sends an S1AP: Initial_Context_Setup_Response message to the MME to acknowledge the successful context setup.
15. The UE sends an RRC: UL_Information_Transfer message to the eNodeB that includes the NAS: Attach_Complete and NAS: Activate_Dedicated_EPS_Bearer_Context_Accept messages.
16. The eNodeB sends the two NAS messages (encapsulated in the S1AP: Uplink_NAS_Transport message) to the MME.
17. After receiving the S1AP: Initial_Context_Setup_Response message and the NAS messages, the MME sends a GTP-C: Modify_Bearer_Request message to the S-GW, indicating the eNodeB user plane address.
18. The S-GW finishes the initial context setup procedure with a GTP-C: Modify_Bearer_Response message to the MME.

The relevant signaling specifications used in this example are [5–10].

9.6 Handover (X2 interface)

Handover (HO) is a procedure in which the UE changes the serving eNodeB. The reason for this change is typically the change in radio conditions caused by the UE moving from one cell area to another. However, there can be other reasons, such as load balancing.

LTE handovers come in many different forms, depending on how the source and the target cells are related: intra-eNodeB HO, intra-MME/S-GW HO via X2, intra-MME/S-GW HO via S1, inter-MME/intra-S1 HO, inter-MME/S1 HO, and inter-RAT HO in various combinations.

In this example, we will show the intra-MME/S-GW HO via the X2 interface. Note that the X2 interface is an optional interface. If the network operator does not support it, then the intra-MME/S-GW HO must be done via the S1 interface. Also, an X2 handover can only be made when both the serving MME and the serving S-GW do not change as a result of the handover (i.e., the source and the target eNodeB has to be served by the same MME and S-GW).

9.6 HANDOVER (X2 INTERFACE)

A handover consists of three different phases:

- Handover preparation;
- Handover execution;
- Handover completion.

These phases are also shown in Figure 9.9.

1. When a data call is ongoing, the UE measures its radio environment according to measurement configuration parameters given to it in RRC: RRCConnectionReconfiguration message. Note that there is no specific measurement request message in LTE, but the measurement configuration is indicated by the MeasConfig IE, which can be attached to the RRCConnectionReconfiguration message. The measurement process is a continuous one; typically, a UE is performing some type of radio measurements periodically while connected to the network.
2. The UE sends back to the source eNodeB (the current serving eNodeB) a RRC: MeasurementReport message. This message can be sent periodically, or it may sent once some triggering event occurs. If the measurement results indicate that there is a more suitable cell available for the UE, the Source eNodeB may trigger a HO.
3. The handover algorithm is operator specific and may include, for example, an X2AP: Resource Status Request message to be sent to a potential target eNodeB in order to find out whether this potential target eNodeB is capable of handling the additional load caused by the UE to be handed over. Note that in E-UTRAN there is no central controller node in the radio access network that could make decisions on which eNodeB is the best base station to handle a UE connection. The source eNodeB has to make this decision itself and may need information from other eNodeBs in this process. However, sending messages to other eNodeBs and waiting for responses takes time, and sometimes HOs have to be made urgently or the call may be lost.
4. If the X2AP: Resource_Status_Response shows that the target eNodeB could handle a new UE connection, the source eNodeB triggers the HO.
5. An X2 handover is started with an X2AP: Handover_Request message from the source eNodeB to the target eNodeB. The message contains information that is needed by the target eNodeB to allocate appropriate resources.

184 PROCEDURES

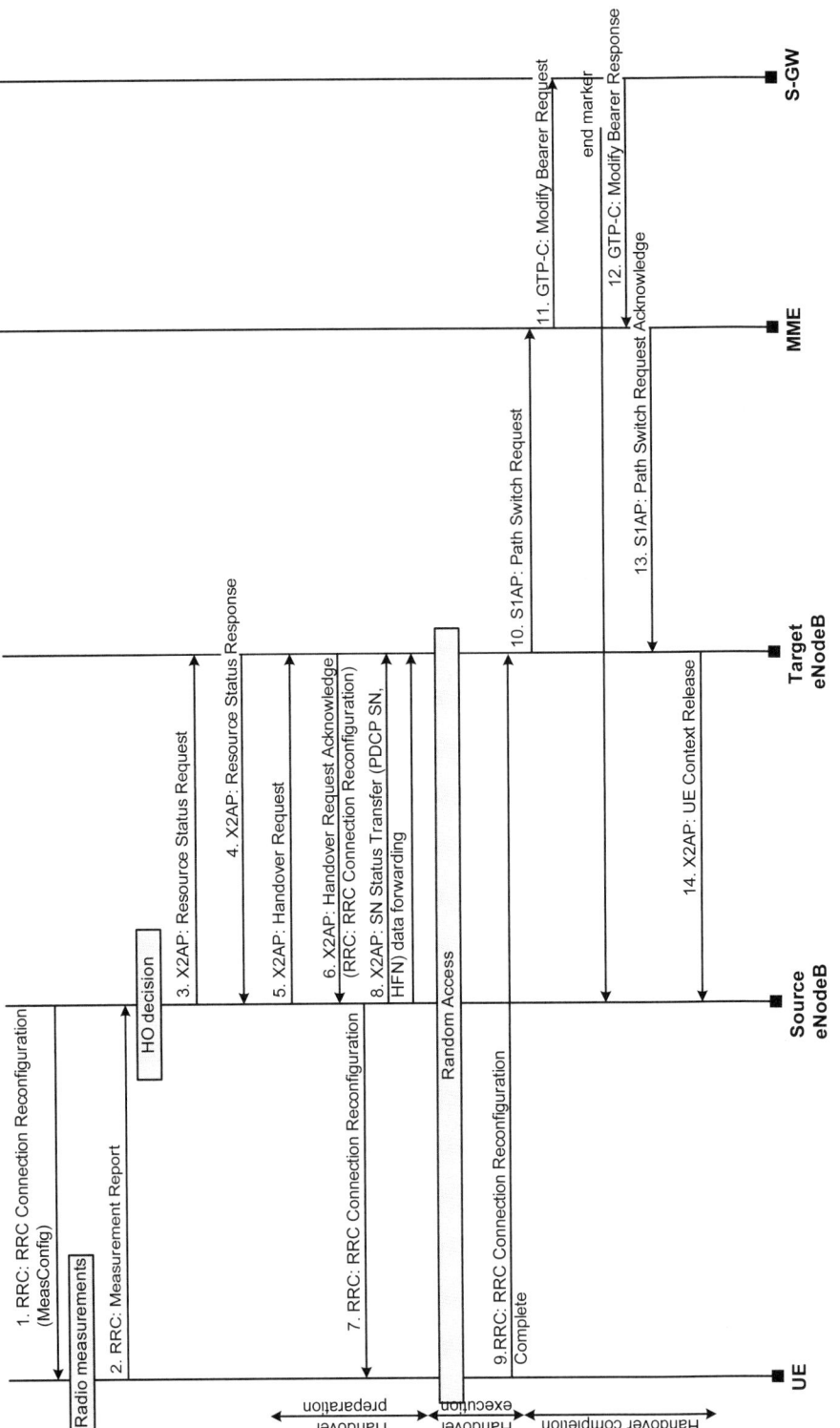

FIGURE 9.9 Handover via the X2 interface.

INTRODUCTION TO 4G MOBILE COMMUNICATIONS

6. The target eNodeB checks the resource availability and if everything is in order, it sends back an X2AP: Handover_Request_Acknowledge message. This message includes a transparent container with a RRC: RRCConnectionReconfiguration message.
7. The source eNodeB unwraps the container and sends the RRC: RRCConnectionReconfiguration message to the UE. This message includes the new C-RNTI, the security algorithm to be used with the target eNodeB, system information, and other information. This is the last message sent by the source eNodeB to the UE.
8. The source eNodeB sends an X2AP: SN_Status_Transfer message to the target eNodeB. It includes the PDCP sequence numbers and the hyper frame numbers (HFN) for the E-RABs. After sending this message, the source eNodeB starts to forward the downlink data packets that it has stored in its buffers (note that the downlink from the source eNodeB to the UE is no longer active).
9. After receiving the RRC: RRCConnectionReconfiguration message, the UE starts a random access procedure toward the target eNodeB. If successful, the UE sends the RRC: RRCConnectionReconfigurationComplete message to the target eNodeB. Now both the uplink and the downlink are active with the target eNodeB.
10. The target eNodeB sends a S1AP: Path_Switch_Request message to the MME to inform it that the UE has changed its serving cell and to request the switch of the downlink GTP tunnel toward a new GTP tunnel endpoint at the target eNB. The MME checks the target eNodeB identity and determines that the existing S-GW can serve also the new eNodeB.
11. The MME sends a GTP-C: Modify_Bearer_Request to the S-GW, to request that the S-GW modify its downlink data bearers (i.e., they should now be addressed to the new target eNodeB).
12. The S-GW starts sending downlink packets to the target eNodeB using the new address and sends the GTP-C: Modify_Bearer_Response to the MME. The S-GW also sends one or more "end marker" packets on the old path toward the S-eNodeB before releasing all user plane resources toward the source eNodeB.
13. The MME sends an S1AP: Path_Switch_Request_Acknowledge message back to the target eNodeB as a notification that the handover was successful.
14. The target eNodeB sends a X2AP: UE_Context_Release message to the source eNodeB indicating that the handover process is complete and that the source eNodeB can now release the resources that were allocated for the UE.

Note that there are many occasions where the handover procedure could have failed. For more information on LTE handovers, including various failing cases, please consult [3, 5, 9].

9.7 CSG Inbound HO

The following example (Figure 9.10) presents a handover from a macro cell to a HeNB. It is shown here because it has some special features when compared to a standard macro-to-macro handover. These include issues such as the closed subscriber group (CSG) and physical cell identity (PCI) confusion.

Home eNodeBs are typically installed by private individuals, and the network operator may not know their exact location, at least not with the accuracy that is needed for efficient network planning, or does not even know if those devices are still there (e.g., the user may have switched his eNodeB off to save electricity at night). That is, when the network sends a Neighbour Cell List to a UE for measurements, it typically does not include HeNBs.

However, the network operator usually knows the approximate location of HeNBs because of emergency call location requirements (possibly reported to the operator with the initial HeNB registration). Therefore, once a UE enters an area where there may be HeNBs that the user can access, the network can send a proximity configuration to the UE, which is basically a request to start monitoring other frequencies too (i.e., frequencies outside the Neighbour Cell List).

1. The procedure starts when the UE has entered an area where there may be HeNBs that the UE can access. These are HeNBs in open or hybrid access modes, or it can be a HeNB in the closed access mode and the UE has a subscription to this CSG. The eNodeB sends a RRC: RRC_Connection_Reconfiguration message to the UE, with the reportProximityConfig field enabled. Because this message is sent in the dedicated mode, it is UE specific. This is a useful feature: when a UE goes around in town, it is typically not configured to monitor and measure closed access mode HeNB cells because that would simply be waste of power. However, when the UE arrives at a macro cell(s) near the user's home (and his HeNB), the network can send a proximity configuration only to this UE via the macro cell so that the UE can search for the HeNB and possibly make a handover to it. Other UEs in the area do not know about this HeNB because their proximity configuration is not active. The reception of proximity configuration triggers an autonomous search procedure (i.e., the UE starts monitoring frequencies other

9.7 CSG INBOUND HO

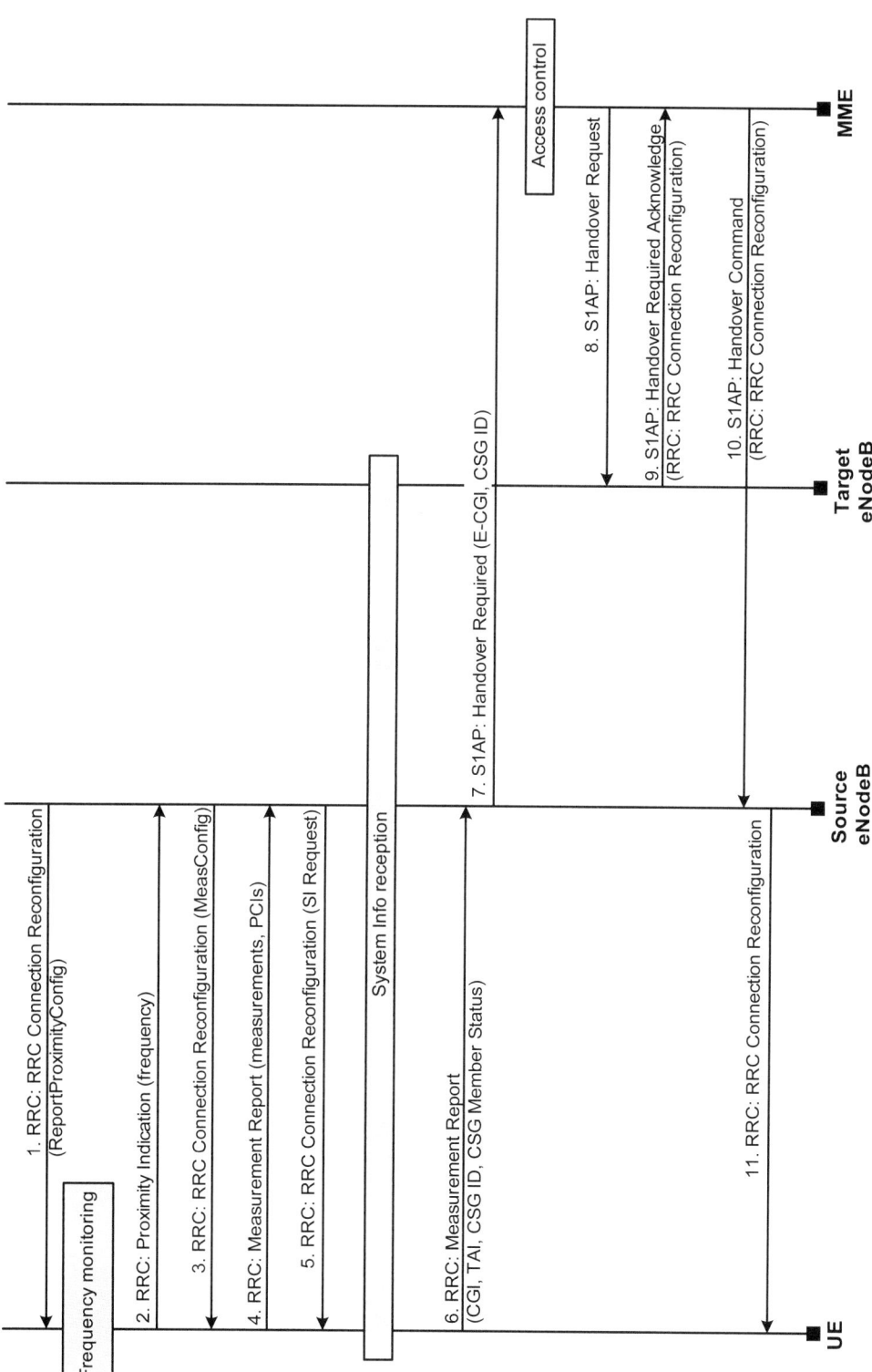

FIGURE 9.10 *Handover to a CSG cell.*

than those in the Neighbour Cell List). The request also indicates which radio acccess technology (RAT) or technologies should be monitored.

2. The UE sends a RRC: Proximity_Indication message when it finds a cell that may be a CSG member cell. The expression is "may" because at this point the UE does not decode SI from the measured channels and cannot know for sure whether or not a cell is a CSG cell. It is not defined how the UE should perform the "autonomous search" for CSG cells (this is left for the UE manufacturer to decide), but most probably it does not mean that the UE should scan all frequencies from all RATs because that would take a very long time (note that this procedure takes place in the connected mode and the UE may be busy communicating with the eNodeB at the same time). Instead, the UE may, for example, measure only those frequencies that it knows belong to CSGs where it is a subscriber. The proximity indication contains the RAT and the frequency of the possible CSG cell.

3. If the eNodeB concludes that one of the indicated cells might be a CSG cell suitable for this UE, it will send a new RRC: RRC_Connection_Reconfiguration message to the UE, including a new measurement configuration that allows the UE to measure the candidate cell regularly, as part of its normal neighbor cell monitoring process.

4. If the candidate CSG cell is better than the current cell or some other event triggers the measurement reporting, the UE sends an RRC: Measurement_Report, which also contains physical cell identities (PCI) of the reported cells.

5. If the received measurement report indicates that a handover could be beneficial, the eNodeB configures the UE to perform SI acquisition, again with a RRC: RRC_Connection_Reconfiguration message. As you may have already gathered, this message is a multipurpose message for the UE connected mode. Whatever needs to be done in the UE, the eNodeB sends this message. It is then the various parameters within the message that actually tell the UE what action should be taken.

6. The UE acquires the SI from the CSG cell and reports the most important parameters back to the eNodeB using an RRC: Measurement_Report message. Note that it is only at this point that the UE finds out if this cell is really a CSG cell, since this piece of information is transmitted in SIB type 1.

7. After receiving the full measurement report from the UE, the eNodeB may decide to trigger a handover. If so, it sends a S1AP: Handover_Required message to the MME, which includes the target E-CGI and the CSG ID, among other parameters. The cell access

mode for the target cell must also be included in case the access mode is hybrid.

8. The MME checks whether this UE is allowed to access the CSG cell in question (i.e., whether the UE is a subscriber of this CSG). If the access control was successful, the MME sends a S1AP: Handover_Request message to the target eNodeB (i.e., the HeNB in this case). If the target cell is a hybrid cell, the CSG membership status will be included in the message.
9. The target HeNB checks that the received CSG ID in the handover request matches the CSG ID of this cell. If so, then the HeNB allocates necessary resources for the UE, and sends back to the MME a S1AP: Handover_Request_Acknowledge message. This message includes a transparent container with a RRC: RRCConnectionReconfiguration message in it.
10. The MME further sends a S1AP: Handover_Command message to the source eNodeB.
11. The source eNodeB unwraps the container and sends the RRC: RRCConnectionReconfiguration message to the UE. This message includes a new C-RNTI, the security algorithm to be used with the target eNodeB, system information, and other information. This is the last message sent by the source eNodeB to the UE.

After this point, the signaling sequence is similar to the previous handover example in Figure 9.9, continuing from point 8.

It is possible or even probable that there is a HeNB gateway between the MME and the HeNB, but it does not have any effect on the signaling procedure in this example.

Note that the handover from a CSG cell to a macro cell follows the normal handover rules; there is no special "CSG" handling in that case. CSG handovers are discussed in [3, 5, 9].

9.8 S1 Release Procedure

S1 Release involves the release of the S1-MME signaling connection, and the release of all UE data bearers from S1-U. Typically this procedure is initiated by the UE switch-off, although there are also several other causes that may lead to S1 release, including various errors or failures in eNodeB or MME (see Figure 9.11).

1. The procedure starts with the eNodeB requesting the UE to release the existing signaling connection with a RRC: RRC_Connection_Release.

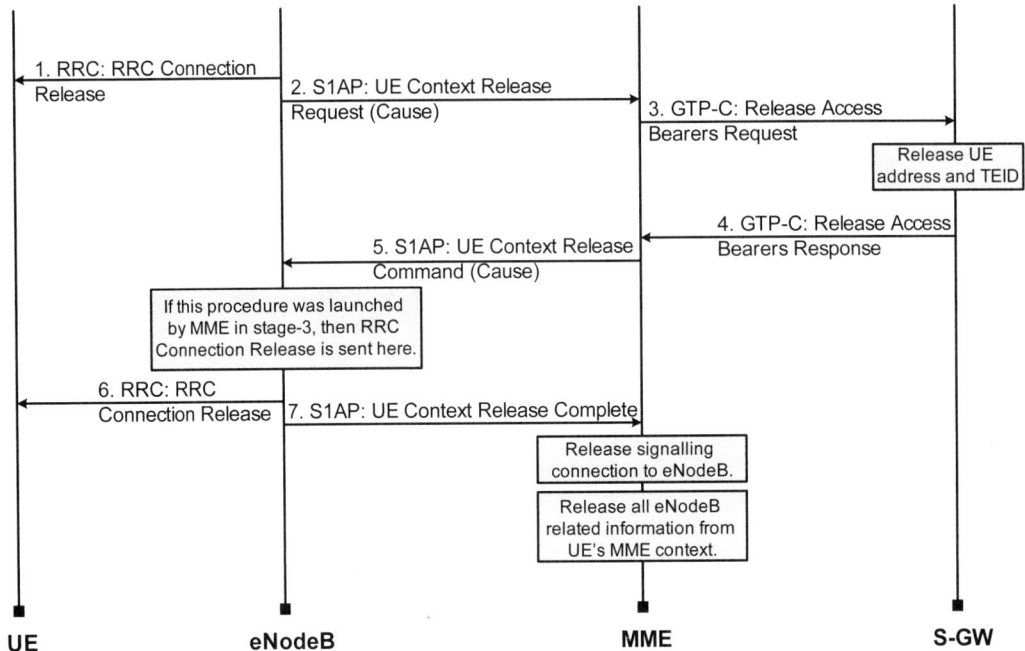

FIGURE 9.11 *S1 release.*

2. The eNodeB sends a S1AP: UE_Context_Release_Request with a cause parameter, which indicates the reason for the release.

3. The MME sends the release request further to the S-GW in GTP-C: Release_Access_Bearers_Request. Note that if the S1 release procedure is triggered by the MME, then this is the first message in the procedure.

4. Upon receiving this message, S-GW releases all eNodeB-related information (address and TEIDs) for the UE and responds with a Release_Access_Bearers_Response message to the MME.

5. MME sends a S1AP: UE_Context_Release_Command (cause) to the eNodeB.

6. If this procedure was launched by the MME in stage 3, then the RRC connection is released here with a RRC: RRC_Connection _Release message.

7. The eNodeB responds to the MME with a S1AP: UE_Context_Release_Complete. Upon receiving this message, the MME releases the signaling connection to the eNodeB and deletes all eNodeB-related information from its stored UE context.

See [3, 5, 7–9] for further information.

9.9 Dedicated Bearer Activation

When the UE attaches to the network, it will be assigned a default EPS bearer. This bearer exists as long as the connection remains active. The UE may also be allocated additional bearers, so called dedicated bearers. Each bearer comes with associated QoS parameters, and the various entities in the network must allocate resources for each new bearer so that they can fulfill the QoS requirements for data that is transported via those bearers. One UE can thus have several dedicated bearers allocated, one for each different data connection.

A dedicated bearer allocation is initiated by the policy and charging rules function (PCRF), although the request for this can come from the UE. In such a case, the signaling flow diagram in Figure 9.12 would be preceded by a NAS: Bearer_Resource_Modification_Request message from the UE to the MME, and thereafter as Bearer_Resource_Command from the MME to the S-GW and further to P-GW

1. The procedure starts when the PCRF sends a PCC decision provision message to the P-GW. This message indicates the QoS that is required for the new bearer.
2. The P-GW transforms the QoS policy requirements into bearer-level QoS parameters and allocates an EPS bearer. The P-GW further sends a GTP-C: Create Bearer Request message to the S-GW.
3. The S-GW in turn allocates the resources for the new bearer and forwards the message to the MME.
4. The MME builds a NAS message (activate dedicated EPS bearer context request) to be sent to the UE, and encapsulates it into S1AP: E-RAB Setup Request message, which is sent to the eNodeB.
5. The eNodeB maps the EPS bearer QoS to the radio bearer QoS. It sends the NAS: Activate Dedicated EPS bearer context request further to the UE, encapsulated into a RRC: RRC_Connection_Reconfiguration message.
6. The UE first acknowledges the new radio bearer activation with a RRC: RRC_Connection_Reconfiguration message.
7. The eNodeB further acknowledges the successful radio bearer activation to the MME with a S1AP: E-RAB_Setup_Response.
8. The UE NAS layer sends a NAS: Activate Dedicated EPS bearer context accept message to the eNodeB within an UL_Information_Transfer message.
9. This message is further relayed to the MME within S1AP: Uplink_NAS_Transport message.
10–11. Thereafter, the successful bearer creation is acknowledged to S-GW and P-GW with a GTP-C: Create_Bearer_Response message.

192 PROCEDURES

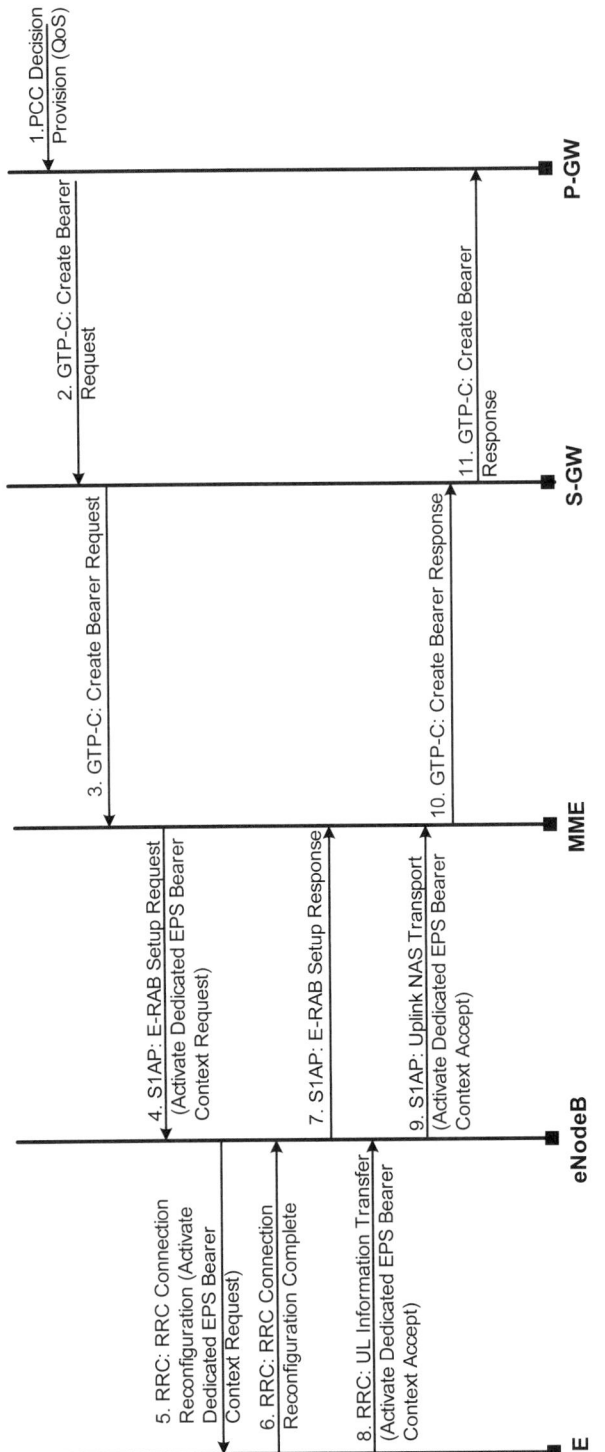

FIGURE 9.12 *Dedicated bearer activation.*

Bearer activation is discussed in many different specifications, for example in [3, 5–9].

REFERENCES

[1] 3GPP TS 36.211, v 11.4.0, Evolved Universal Terrestrial Radio Access (E-UTRA); Physical Channels and Modulation, 09/2013.

[2] 3GPP TS 36.321, v 11.3.0, Evolved Universal Terrestrial Radio Access (E-UTRA); Medium Access Control (MAC) Protocol Specification; 06/2013.

[3] 3GPP TS 36.300, v 11.7.0, Evolved Universal Terrestrial Radio Access (E-UTRA) and Evolved Universal Terrestrial Radio Access Network (E-UTRAN); Overall Description; Stage 2; 09/2013.

[4] 3GPP TS 36.213, v 11.4.0, Evolved Universal Terrestrial Radio Access (E-UTRA); Physical Layer Procedures; 09/2013.

[5] 3GPP TS 36.331, v 11.5.0, Evolved Universal Terrestrial Radio Access (E-UTRA); Radio Resource Control (RRC); Protocol Specification; 09/2013.

[6] 3GPP TS 24.301, v 12.2.0, Non-Access-Stratum (NAS) Protocol for Evolved Packet System (EPS); Stage 3; 09/2013.

[7] 3GPP TS 36.413, v 11.5.0, Evolved Universal Terrestrial Radio Access Network (E-UTRAN); S1 Application Protocol (S1AP); 09/2013.

[8] 3GPP TS 29.274, v 12.2.0, 3GPP Evolved Packet System (EPS); Evolved General Packet Radio Service (GPRS) Tunneling Protocol for Control Plane (GTPv2-C); Stage 3; 09/2013.

[9] 3GPP TS 23.401, v 12.2.0, General Packet Radio Service (GPRS) Enhancements for Evolved Universal Terrestrial Radio Access Network (E-UTRAN) Access; 09/2013.

[10] 3GPP TS 29.272, v 12.2.0, Evolved Packet System (EPS); Mobility Management Entity (MME) and Serving GPRS Support Node (SGSN) Related Interfaces Based on Diameter Protocol; 09/2013.

CHAPTER 10

Specifications

10.1 Introduction

This chapter section discusses how 3GPP standards are developed; the structure of 3GPP as an organization; its various working groups, working practices, releases; and more. The purpose of this chapter is to familiarize the reader with 3GPP and especially its specifications. The specifications are the ultimate source of LTE-A information; all technical information about the system is there. The only question is whether it can be found and understood.

3GPP was a rather unique arrangement when it was set up in 1998. It is a collaboration project between regional telecommunications standards organizations. Currently 3GPP has six organizational partners: ARIB, ATIS, CCSA, ETSI, TTA, and TTC. Individual companies can join 3GPP via their organizational partner association (i.e., a company cannot join "only" 3GPP without first joining one of the organizational partners). As of October 2013, there are 401 individual 3GPP members. This group includes equipment vendors, network operators, governmental organizations, regulators, academic establishments, research laboratories, and so on (see Table 10.1).

Also, at this point it is good to remind the reader what the 3GPP specifications actually specify. Typically they specify the functionality of *interfaces*. For example, there is no 3GPP specification that specifies how the eNodeB should be implemented. Instead, several specifications specify how the interfaces around the eNodeB work, as well as what information and in what format should be transported over those interfaces. The 3GPP specifications do not want to stipulate or restrict the implementation of eNodeBs (or any other 3GPP network entities for that matter). But if the interfaces are properly specified, it will be possible to use hardware from different vendors on both sides of an interface.

10.2 Internal Structure

The internal structure of 3GPP is depicted in Figure 10.1.

SPECIFICATIONS

TABLE 10.1 List of 3GPP Organizational Partners

Abbreviation	Full Name	Website
ARIB	The Association of Radio Industries and Businesses	www.arib.co.jp
ATIS	The Alliance for Telecommunications Industry Solutions	www.atis.org
CCSA	China Communications Standards Association	www.ccsa.org.cn
ETSI	The European Telecommunications Standards Institute	www.etsi.org
TTA	Telecommunications Technology Association	www.tta.or.kr
TTC	Telecommunication Technology Committee	www.ttc.or.jp

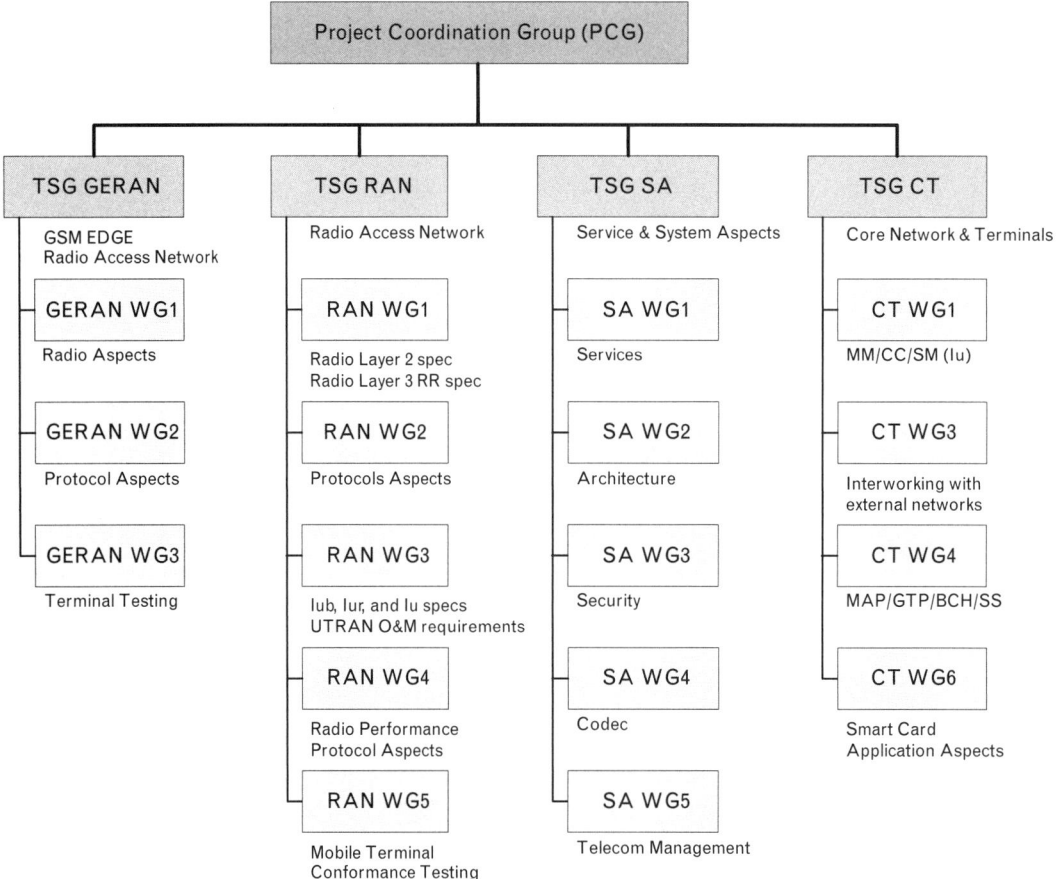

FIGURE 10.1 *3GPP structure.*

On top of the structure is the Project Coordination Group (PCG). The PCG is the highest decision making body in 3GPP. It meets formally every six months to carry out the final adoption of 3GPP Technical Specification Group work items, to ratify election results, and determine the resources committed to 3GPP. PCG delegates include the leadership from

3GPP technical specification groups (TSG), delegations from 3GPP organizational partners, and also representatives from market representation partners (MRP).

The actual technical work is done in TSGs. Currently there are four of them:

- Radio access networks (RAN);
- Service and systems aspects (SA);
- Core network and terminals (CT);
- GSM EDGE radio access networks (GERAN).

Furthermore, the work in TSG is organized into working groups (WG). Each TSG can have several WGs reporting to it. Each WG is responsible for certain aspects of the TSG work. WGs meet four to six times a year, usually in five-day-long meetings. Occasionally some difficult issues may result in additional ad hoc meetings. WG meetings are very technical in nature. WGs can and do propose changes to specifications via a formal process called change requests. However, CRs agreed in WGs still have to be submitted to the next TSG meeting for the final approval.

TSGs currently meet four times a year. In TSG meetings, the results from the work of WGs is presented for information, discussion, and approval. Only after a TSG has approved a CR can it be incorporated in specifications. In following sections, we will discuss each of the TSGs in turn, as well as introduce their WGs.

10.2.1 TSG RAN

The TSG RAN is responsible for the UTRAN and evolved UTRAN, including their internal structures and functions, of systems for evolved 3G and beyond. Specifically, it has a responsibility for the followinbg:

- Specification of layer 1 of Uu radio interface for UE, Node B, and eNode B, as well as layer 1 of Un radio interface for relay node and eNode B;
- Specification of layer 2 of Uu radio interface for UE, Node B, and eNode B, as well as layer 2 of Un radio interface for relay node and eNode B;
- Specification of layer 3 of Uu radio interface for UE, Node B, and eNode B, as well as layer 3 of Un radio interface for relay node and eNode B;

- Overall UTRAN and E-UTRAN architecture, including the specification of interfaces between UTRAN/E-UTRAN and the core network, and of network interfaces within UTRAN/E-UTRAN;

- Specification of LTE positioning protocol (LPP) between the UE and the positioning server for LTE, and LTE positioning protocol A (LPPa) between the eNode B and the positioning server;

- UTRAN and evolved UTRAN O&M requirements;

- Transport of implementation-specific O&M between the management system and node B;

- Conformance test specifications for evolved UE;

- Conformance test specifications for base stations and evolved base stations;

- Specifications for radio performance and RF system aspects.

TSG RAN has five WGs. These are large working groups; some of them have 200–300 delegates attending every meeting. RAN WG1 (radio layer 1) is responsible for the specification of the physical layer of the radio interface for UE, UTRAN, evolved UTRAN, and beyond, covering both FDD and TDD modes of the radio interface. RAN1 is usually the largest 3GPP WG in terms of delegate numbers.

RAN WG2 (radio layer 2 and radio layer 3 RR) is in charge of the radio interface architecture and protocols (MAC, RLC, PDCP), the specification of the radio resource control protocol, the strategies of radio resource management, and the services provided by the physical layer to the upper layers. Within the scope of TSG RAN, RAN WG2 is responsible for the development of specifications dealing with UTRA, evolved UTRA, and beyond.

RAN WG3 is responsible for the overall UTRAN/E-UTRAN architecture and the specification of protocols for the Iu, Iur, Iub, S1, and X2 interfaces.

RAN WG4 works on the RF aspects of UTRAN/E-UTRAN. RAN WG4 performs simulations of diverse RF system scenarios and derives the minimum requirements for transmission and reception parameters and for channel demodulation. Once these requirements are set, the group defines the test procedures that will be used to verify them (only for BS). Requirements for other radio elements, like repeaters, are specified in the RAN WG4 as well.

RAN WG5 works on the specification of conformance testing at the radio interface (Uu) for the user equipment (UE). The test specifications are based on the requirements defined by other groups such as RAN WG4 for the radio test cases, and RAN WG2 and CT WG1 for the signaling and protocols test cases. RAN WG5 is organized in two subgroups, the RF subgroup and the signaling subgroup.

10.2.2 TSG SA

The TSG service and system aspects (TSG-SA) group is responsible for the overall architecture and service capabilities of systems based on 3GPP specifications and, as such, has a responsibility for cross TSG coordination. That is, even though all TSGs are equal, TSG-SA is more equal than the others. This position is also reflected on the fact that TSG-SA has its plenary meeting after other TSG meetings, and other TSGs will present their meeting reports to TSG SA.

TSG SA has also five WGs. These groups have quite diverse tasks. SA WG1 handles services. This group is typically the first one to have a look at new features and services that might be added to the system. The outputs of this working group are technical specifications and reports, or changes to these, which are all submitted to TSG SA for approval. Once approved, they form the basis for the work for the whole of 3GPP and for industry segments interested in deploying networks based on IMS. That is, in a way TSG SA1 works as a filter. It examines which features or services are worth adding to the standards and then produces requirement specifications (known as stage 1 specifications) that will be used by other 3GPP groups as guidance in their work. This also means that while other 3GPP groups are preparing specifications for release X, TSG SA1 may already be discussing release X+1 features.

SA WG2 handles the system architecture. It is in charge of developing the stage 2 specifications for the 3GPP network. Based on the service requirements elaborated by SA WG1, SA WG2 identifies the main functions and entities of the network, how these entities are linked to each other, and the information they exchange. The output of SA WG2 is used as input by other groups in charge of the definition of the precise format of messages in stage 3 specifications (stage 2 for the radio access network is under TSG RAN's responsibility). The group has a systemwide view and decides how new functions integrate with the existing network entities. Therefore, SA WG2 is probably the most important WG in the whole 3GPP.

SA WG3 is responsible for security and privacy in 3GPP systems, determining the security and privacy requirements, and specifying the security architectures and protocols. This WG also ensures the availability of cryptographic algorithms, which need to be part of the specifications. The sub-WG SA3-LI provides the requirements and specifications for lawful interception in 3GPP systems.

SA WG4 handles various codecs in the system. It deals with the specifications for speech, audio, video, and multimedia codecs, in both circuit-switched and packet-switched environments. Other topics within the mandate of SA WG4 are quality evaluation, end-to-end performance, and interoperability aspects with existing mobile and fixed networks from codec point of view.

SA WG5 specifies the requirements, architecture, and solutions for provisioning and management of the network (RAN, CN, IMS) and its services. The WG will define charging solutions in alignment with the related charging requirements developed by relevant WGs and will specify the architecture and protocols for charging of the network and its services.

The WG will ensure its work is also applicable to the management and charging of converged networks and potentially applicable to fixed networks.

10.2.3 TSG CT

TSG CT was formed in 2005 as a merger of two older TSGs, the core network (CN) and terminal (T). TSG CT is responsible for specifying the CN and T equipment aspects of systems based on 3GPP specifications. This includes the following:

- User equipment–core network layer 3 protocols in CS and PS domain excluding the radio access technology layers;
- Core network internal interfaces for call-associated and noncall-associated signaling;
- Interconnection of the core network with external networks;
- SIM/USIM/ISIM/HPSIM and its interface specifications;
- Terminal- or network-based applications supported by 3GPP terminals;
- Core network protocols for CS and PS domain;
- IP multimedia subsystem.

After the merger, it had six WGs, though two of them have been since closed.

CT WG1 is responsible for specifications that define the user equipment–core network l3 radio protocols and core network side of the Iu reference point. Specifically, it has a responsibility for user equipment–core network layer 3 radio protocols (call control, session management, mobility management, SMS).

CT WG2 does not exist. The old T WG2 was responsible for terminal capabilities, but that group was closed in 2005.

CT WG3 specifies the bearer capabilities for circuit and packet switched data services, as well as the necessary interworking functions toward both the user equipment in the UMTS/LTE PLMN and the terminal equipment in the external network. In addition CT3 is responsible for end-to-end QoS in the UMTS core network.

CT WG4 standardizes stage 2 and stage 3 aspects within the core network focusing on supplementary services, basic call processing, mobility management within the core network, bearer-independent architecture, GPRS between network entities, transcoder free operation, CAMEL, generic user profile, wireless LAN–UMTS interworking, and descriptions of IP multimedia subsystem. CT WG4 is also responsible as a "protocol steward" for the some IP-related protocols (this involves analyzing, validating, extending if necessary, clarifying usage, specifying packages, and parameter values).

CT WG5 used to develop application programming interfaces (APIs) for the open service access (OSA). However, this task was transferred to the Open Mobile Alliance (OMA) in 2008 and the WG was closed.

CT WG6 is responsible for the development and maintenance of specifications and associated test specifications for 3GPP smart card applications and the interface with the mobile terminal. These include the following:

- Subscriber identity module (SIM), which is used by 2G systems;
- Universal subscriber identity module (USIM), which is used by 3GPP systems;
- IM services identity module (ISIM), with the exception of the security algorithms (those are developed by SA WG3).

10.2.4 TSG GERAN

TSG GERAN is responsible for the specification of the radio access part of GSM/EDGE. Since 1992, GSM specifications had been developed by ETSI, but in 2000 this work was transferred to a new GERAN group within the 3GPP. Since GERAN still retains some old working practices, its meeting schedule and working practices are different from other TSGs. For example, GERAN TSG and WG meetings take place during the same week at the same location; typically Monday and Friday are reserved for the TSG plenary and Tuesday through Thursday for working group meetings. GSM/EDGE has a long and distinguished history, but now it seems that operators are no longer interested in investing in new GERAN development work and are moving their focus to 3G/4G. Subsequently, the amount of work in TSG GERAN has diminished, and it is possible that this TSG will disappear in not so distant future.

- Currently TSG GERAN has three WGs. GERAN WG1 is responsible for the following:

- RF aspects of GERAN;
- Internal GERAN interface specifications such as A-ter (CCU-TRAU);
- Conformance test specifications for testing of all aspects of GERAN base stations;
- GERAN-specific O&M specifications for the nodes in the GERAN;

GERAN WG2 is responsible for the protocol aspects of GERAN. GERAN2 specifies the data link and RLC/MAC layer protocols and the interfaces between these layers and the physical layer.

GERAN WG3 is responsible for conformance test specifications for testing of all aspects of GERAN terminals. It also deals with GERAN radio aspects and interfaces.

10.2.5 Mobile Competence Center (MCC)

In addition to TSGs and WGs, 3GPP also includes the mobile competence center (MCC). MCC provides support to the Third Generation Partnership Project (3GPP). It is about 20 people strong and is located at the ETSI headquarters in Sophia Antipolis in southern France. Each TSG/WG has a dedicated MCC support officer who gives support at 3GPP meetings, takes minutes, implements agrees actions, updates specifications, and so on.

10.3 Standardization Process

10.3.1 Introduction

Even if the reader is not planning to participate in 3GPP standardization, it is good to know how the standardization is done. The process is not obvious for an outsider, and the jargon used may not be self-explanatory. Understanding the process will help to find the right information from the specifications.

In the previous section the various TSGs and WGs were discussed. The standardization work is done in these groups, either in face-to-face meetings, via email lists, or in electronic/telephone meetings. 3GPP standardization work is contributions-driven. It is up to individual member companies to propose new work items or changes to specifications.

10.3.2 Work Items

In principle, all work done in 3GPP should belong to some work item. A work item is a piece of work introducing a new feature, modifying an existing feature, or perhaps studying some proposed new feature. It must be initiated and supported by a group of companies (at least four companies by current rules). Work item proposals must be approved by the relevant TSG before any related work can commence. If a work item proposal is approved, it will become part of the work plan and may get meeting time from relevant WGs.

In its specification work, 3GPP uses a three-stage methodology as defined in ITU-T recommendation I.130 [1]. That is, there are three types of specifications, and, at least in principle, a new feature to be standardized should be defined in all of these:

- Stage 1: these specifications contain the service requirements from the user point of view.
- Stage 2: these specifications define the architecture to support the service requirements.
- Stage 3: these specifications define an implementation of the architecture by specifying protocols in details.

However, in practice a new feature, especially if it is a minor enhancement, may only include stage 2 and/or stage 3 specification changes.

If the work item is a so called study item, making a feasibility study on whether a new feature is needed, then its outcome is recorded in a technical report, which could be regarded as a stage 0 specification. Technical reports are not specifications as such; they merely contain a description of a feature and say whether it is feasible to include it in normative 3GPP specifications. If the conclusion is yes, then a related work item can be proposed. Despite this, note that technical reports are often very useful sources of information for a reader who wants to learn about a new 3GPP feature. The actual technical specifications are compact standards that do not explain issues. However, the corresponding technical report, if such exists, may help to understand the topic in hand. But it is also good to remember that technical reports describe only the study phase. There is no guarantee that the actual technical specification will follow the conclusions and recommendations of the technical report. In any case, technical reports are typically not maintained after the study phase has concluded, and thus they should never be used as an aid for the implementation of the feature.

Each work item has a rapporteur. This person is typically from the company that initiated the WI proposal and is an expert in the subject. When the specification is still in a draft phase, the rapporteur collects the agreed text proposals toward the draft specification at each meeting and updates the draft specification accordingly. Once the specification is considered to be 80 percent ready, it will be put under change control and its maintenance becomes the responsibility of the MCC. Thereafter, the treatment of the specification becomes more formal.

Changes to a specification under change control can only be introduced via formal change requests (CRs), see Figure 10.2. A CR must be first *agreed* by the WG, and then *approved* by the corresponding TSG. Only after that will MCC incorporate the change into the specification. A CR can introduce a new feature, modify an existing feature, or fix an error that was found from existing definitions. Also, editorial corrections are introduced via CRs. Each CR can modify only a single version of a single specification. For example, if the modification of a feature affects several specifications, then each of those specifications needs its own CR. A CR must clearly show what should be changed in the specification and provide reasons for the change. All approved CRs will be recorded in the central CR database, and they are also listed in the end of the relevant specification.

10.3.3 Version Numbering

Specifications are updated after every TSG plenary meeting by the MCC, provided that the TSG approved any CRs for the specification in question. Since currently there are four meetings per year for each TSG, it follows that

FIGURE 10.2 *Example of a CR cover page.*

```
3GPP TSG-??? Meeting #nn                                          DocNumber
Location, Country, Date

                                                                  CR-Form-v11
                        CHANGE REQUEST
    SpecNumber    CR  CRNum    rev  -    Current version:   x.y.z

         For HELP on using this form: comprehensive instructions can be found at
                     http://www.3gpp.org/Change-Requests.

    Proposed change affects:    UICC apps☐   ME☐  Radio Access Network☐  Core Network☐

    Title:
    Source to WG:
    Source to TSG:
    Work item code:                                  Date:    yyyy-MM-dd
    Category:                                        Release: Rel-
                Use one of the following categories:     Use one of the following releases:
                   F (correction)                         Rel-4   (Release 4)
                   A (mirror corresponding to a change in an earlier
                      release)                            Rel-5   (Release 5)
                   B (addition of feature),               Rel-6   (Release 6)
                   C (functional modification of feature) Rel-7   (Release 7)
                   D (editorial modification)             Rel-8   (Release 8)
                Detailed explanations of the above categories can  Rel-9   (Release 9)
                be found in 3GPP TR 21.900.               Rel-10  (Release 10)
                                                          Rel-11  (Release 11)
                                                          Rel-12  (Release 12)
                                                          Rel-13  (Release 13)

    Reason for change:
    Summary of change:
    Consequences if not
    approved:
    Clauses affected:
                           Y N
    Other specs                 Other core specifications    TS/TR ... CR ...
    affected:                   Test specifications          TS/TR ... CR ...
    (show related CRs)          O&M Specifications           TS/TR ... CR ...
    Other comments:
```

the specifications will be updated a maximum of four times a year. Once a specification is updated, it will be given a new version number. The version number follows an x.y.z notation, where:

- x gives the release number. Release numbers 0, 1, and 2 indicate that the specification is a draft that is not yet approved by a TSG. Once the specification is approved (i.e., it is estimated to be 80 percent ready), the specification is given version number x.0.0. For example if this is a Release-11 specification, then the new version number will be 11.0.0.

- y gives the technical version number. For the first version of a release, y is always 0. The version number is incremented by one each time a technical change (or changes) is introduced to the specification. Each TSG meeting that approves one or more CRs will make such an increment. The new version will contain all approved CRs from a TSG meeting.

- z gives the editorial version number. An editorial change is one that does not affect the technical content of the specification. These can include correction of typos or a clarification of the meaning of the specification. The z field is reset back to zero every time the y field is changed.

10.3.4 Releases

A release forms a self-contained set of specifications that can be used to construct a functional 3GPP system. Specifications for release X can be created by taking the corresponding Release (X-1) version as a starting point and then adding all approved release X CRs to it. Once a release is considered mature enough, it will be frozen. Frozen in this context means functional freeze: no new functionality will be added to a frozen release, but errors found can be corrected. The term frozen release has caused misunderstandings in the past, and it must be stressed that a frozen release does not mean that the specifications in that release remain unchanged. A company can propose CRs to a frozen release as long as the release remains open. Once a release is declared closed, it is no longer maintained by the MCC. Currently, closed releases include all releases up until and including Release 99.

However, even though CRs can be proposed against old but open releases, the chances of getting a CR approved becomes progressively more difficult with time (i.e., the member companies are very reluctant to re-open very old releases). It has to be a very serious error found from, for example, a specification that was frozen five years ago before a CR correcting the error can get an approval in the TSG. The reason member companies often do not approve such changes is quite practical: if a change were to be introduced in old specifications, it might result, at least in theory, in product recalls to fix the problem in products that are already out there, and with mass-market devices such as mobile phones this would be a completely impossible situation.

It can also be argued that a CR fixing an old error is unnecessary on the grounds that the system with the error in it has been in use for a long time already, and therefore the problem cannot be serious and the specifications can be left untouched.

The freezing date for a release under work is set well in advance. However, it is not unheard of that the freezing date is postponed if the work takes more time than expected and important issues remain open in the release.

Earlier 3GPP releases up until Release 7 were briefly discussed earlier in this book in the history chapter. From Release 8 onward the releases contain an LTE component, and from Release 10 onward they also contain LTE-advanced items. We will discuss the main enhancements introduced in these releases in this section. Note that the freeze dates given are freeze dates for stage 3 specifications. Stage 1 and stage 2 specifications may have their own freezing dates, which are typically much earlier than stage 3 freezing dates. See Figure 10.3 for the LET release schedule.

10.3.4.1 Release 8

Release 8 introduced LTE—including the new air interface and the new type of core network. These were major changes. It was decided that the new air interface definitions cannot fit into existing specifications, and thus a new set of standards, the 36-series, was established. Currently, this series has more than 150 technical specifications and technical reports. Since Release 8 employs a new air interface technology, OFDMA, it cannot be backwards compatible with Release 7. OFDMA technology is discussed in detail in Chapter 4. Other LTE-specific Release 8 features include:

- Flat RAN architecture, RAN having only a single type node (eNodeB);
- A new core network, enhanced packet core (EPC);
- Low latency (transit time < 10ms);
- MIMO support (max 4x4 downlink, 2x2 uplink);
- QPSK/16QAM/64 QAM modulation;

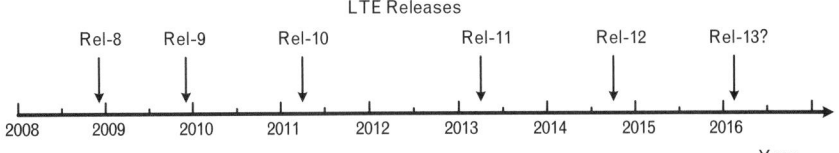

FIGURE 10.3
LTE release schedule.

Note: The Release publication date given here is the Stage-3 functional freeze date.

- Multicast/broadcast over a single frequency network (MBSFN);
- Data rates up to 300-Mbps downlink, 75-Mbps uplink.

Note that LTE air interface is also known as evolved UMTS terrestrial radio access (E-UTRA). Release 8 was frozen in December 2008.

10.3.4.2 Release 9

Release 9, by nature, is a smaller "enhancement" release in between Release 8, which introduced LTE, and Release 10, which introduced LTE-advanced. However, Release 9 still contains many important new features. These include:

- Home eNodeBs (i.e., femtocells);
- Evolved multimedia broadcast and multicast service (eMBMS);
- The self-organizing network (SON);
- Location services (LCS) to detect the location of a mobile device accurately;
- Multistandard radio (MSR).

Home eNodeBs are further discussed in Section 7.3, eMBMS in Section 11.6, and SON in Section 11.7. Release 9 was frozen in December 2009.

10.3.4.3 Release 10

Release 10 introduced LTE-advanced, which is again a major release with greatly enhanced capabilities. LTE-A increases the capacity and throughput of the LTE system. UEs near cell edges will also see improved performance as a result of Release 10. The main features include:

- Data rates up to 3-Gbps downlink and 1.5-Gbps uplink;
- Carrier aggregation (CA), allowing the combination of up to five separate carriers together, in theory providing bandwidths up to 100 MHz. However, in Releases 10 and 11, the maximum number of component carriers is two;
- Higher order MIMO: 8×8 downlink and 4×4 uplink;
- Relay nodes;
- Enhanced intercell interference coordination (eICIC) to improve performance at cell edges;

- SON enhancements;
- Minimization of drive tests feature;
- HeNB mobility enhancements;
- Machine type communications (MTC).

Release 10 was frozen in March 2011.

10.3.4.4 Release 11

Release 11 was again an enhancements release after bigger changes introduced in Release 10. The list of refinements include:

- Enhancements to carrier aggregation;
- New frequency bands;
- Coordinated multipoint (CoMP) transmission and reception (simultaneous communication with multiple cells);
- Network-based positioning support (NBPS);
- Network energy saving;
- Further enhanced intercell interference coordination (FeICIC).

Release 11 was frozen in March 2013. Note that the official freeze date was September 2012, but protocols (especially RAN protocols) were not stable until six months later.

10.3.4.5 Release 12

Potential features for Release 12 were discussed at a 3GPP workshop in Slovenia in June 2012. A strong requirement was the need to support the rapid increase in mobile data usage, but other items included the efficient support of diverse applications while ensuring a high-quality user experience.

At the time of this writing, Release 12 work is still at early phases, and the list of features in it may still change. Currently Release 12 seems to introduce at least:

- Enhanced LIPA and SIPTO;
- MTC enhancements;
- Proximity-based services;
- Enhancements to HeNB mobility;

10.3 STANDARDIZATION PROCESS

- Support for BeiDou satellite navigation;
- Improved mobility for heterogenous networks (HetNet) (i.e., networks with macro, pico, and femto cells);
- 8x8 MIMO for uplink.

Release 12 (stage 3) is expected to be frozen in September 2014.

10.3.5 Development Cycle

The development cycle for specifications is shown in Figure 10.4. Once a release is frozen, no new functionality will be added to it, but corrections will be accepted. The focus of the specification work will move to the next release. As soon as there are approved CRs for the new release X, a copy of the latest release X-1 specification is named the release X specification, and the changes from the CR will be added to it.

CRs will be handled and agreed in WG meetings and then given the final approval in TSG meetings. In special cases it is also possible for the companies to submit their CRs directly to a TSG meeting. After a TSG meeting, all approved CRs are bundled together per specification, and then the specification is updated. Note that a correction may affect several releases of the same specification. However, in this case each release requires a CR of its own. Between each TSG meeting there can be one or more WG meetings. Since there are four TSG meetings in a year, the specifications are updated at most four times a year. If there are no approved CRs for a given specification, then it will remain untouched.

Once the release X is frozen, release X specifications must form a self-contained set of standards. If there have been no release-X CRs against a certain specification, and this specification is needed in release X, at this

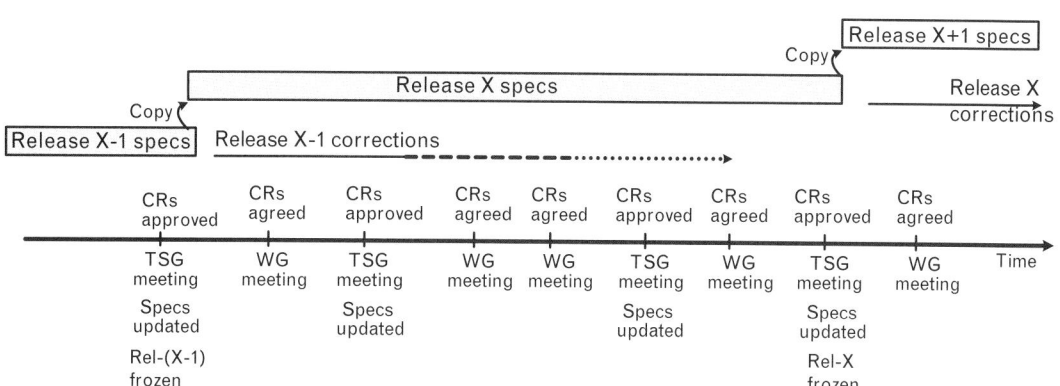

FIGURE 10.4 *The development cycle for specifications.*

point a copy of Release (X-1) specification is taken, and it is renamed the Release X specification.

10.4 Specification Numbering

3GPP specifications are divided into several series. The series numbering goes from 0 to 55, although not all numbers in this range are used. Table 10.2 shows the different series and a short description of their contents.

In the context of this book, we are mostly interested in 36 series, which contains the radio specifications for LTE and LTE-A. However, it is good to know that 36 series is not the only place that contains LTE/LTE-A definitions. The 36 series is about radio access technology only. Many other issues, such as security, charging, and codecs, are in their own series, and often the same specification contains definitions for GERAN, UTRAN, and EU-TRAN systems. Moreover, EPS definitions for the evolved core network

TABLE 10.2   3GPP Specification Series

Series Description	R99 and Later	GSM (Release 4 and Later)	GSM (Before Release 4)
General information (long defunct)			00 series
Requirements	21 series	41 series	01 series
Service aspects ("stage 1")	22 series	42 series	02 series
Technical realization ("stage 2")	23 series	43 series	03 series
Signaling protocols ("stage 3")—user equipment to network	24 series	44 series	04 series
Radio aspects	25 series	45 series	05 series
CODECs	26 series	46 series	06 series
Data	27 series	47 series (none exists)	07 series
Signaling protocols ("stage 3")—(RSS-CN) and OAM&P and charging (overflow from 32 range)	28 series	48 series	08 series
Signaling protocols ("stage 3")—intrafixed-network	29 series	49 series	09 series
Program management	30 series	50 series	10 series
Subscriber identity module (SIM / USIM), IC cards, test specs	31 series	51 series	11 series
OAM&P and charging	32 series	52 series	12 series
Access requirements and test specifications	—	13 series	13 series
Security aspects	33 series	—	—
UE and (U)SIM test specifications	34 series	—	11 series
Security algorithms	35 series	55 series	—
LTE (evolved UTRA) and LTE-advanced radio technology	36 series	—	—
Multiple radio access technology aspects	37 series	—	—

are added to 22, 23, and 24 series, often embedded to existing specifications. However, this section discusses the 36 series only.

Since LTE and LTE-A define the new radio access network, EUTRAN, the 36 series specifications are drafted and maintained by TSG RAN and its WGs. This series currently contains about 150 specifications, and the number is increasing. Not all of these documents are nominative specifications; there are also so called technical reports (TR) in the 36 series.

There are two types of technical reports:

- Those intended to be transposed and issued by the organizational partners as their own publications. These type of TRs are given numbers of the form xx.9xx.

- Those that are 3GPP internal working documents and not intended for external publication. These may include, for example, results of feasibility studies. They are numbered as xx.8xx.

The rest of the series is technical specifications (TS). They are subdivided by the working group responsible as follows:

- 36.1xx: RAN4 specifications (i.e., radio performance related specifications);
- 36.2xx: RAN1 specifications (i.e., air interface physical layer specifications);
- 36.3xx: RAN2 specifications (i.e., air interface protocol specifications);
- 36.4xx: RAN3 specifications (i.e., radio access network protocol specifications);
- 36.5xx: RAN5 specifications (i.e., radio test specifications).

10.5 Backwards Compatibility

Backwards compatibility is an important basic principle that is maintained whenever possible while introducing changes to specifications. It means that changes should be introduced in a way that allows hardware or software to conform to an older version of the specification and to continue functioning without problems in a network that has devices conforming to the new version of the specification. For example, a Release 8 UE should work without problems with a Release 10 eNodeB. Of course, a new feature or a modification of functionality introduced by this new change will not be available for the older equipment, but the change should be done in a way that enables the older equipment to ignore the newer functionality without problems.

Backwards compatibility has important practical reasons behind it. If a customer buys a new LTE mobile phone, it should continue working in any LTE network, regardless of the particular release the network is conforming to. Also, a network operator probably wants to upgrade its network in phases. It is not practical for an operator to upgrade all devices in a network simultaneously, which would be the case without backwards compatibility. If the releases are backwards compatible, then devices confirming to different releases can coexist in the same network.

The backwards compatibility issue does have implications for specifications under work. A backwards compatible solution may not be the most elegant one or the most obvious one. But the principle of backwards compatibility is more important than specification's style points. Note that sometimes backwards compatibility can be broken on purpose. For example, if a serious error that has broken the backwards compatibility in the first place is found, then it is allowed to fix this error with a non–backwards compatible change. The reasoning here is that the original feature would not have worked anyway, so there is no point in trying to make its correction backwards compatible with something that does not work. However, if there is an even older version of the same piece of specification (i.e., older than the version that introduced the error), then it probably a good idea to make the new corrected specification backwards compatible with the old specification.

Backwards compatibility has to be kept in mind even when introducing new features that do not exist in older versions. They have to be introduced in a form that makes it easy to modify them later—especially in a way that is backwards compatible with the original feature.

10.6 E-UTRAN Specifications

This section briefly discusses the most common LTE interfaces and specifications in them. Figure 10.5 presents Uu, S1, and X2 interfaces and the corresponding specifications.

Note that both S1 and X2 interfaces are wired interfaces, and thus the lower layer specifications for those interfaces are not defined by 3GPP. Rather, the specifications listed in Figure 10.5 contain references to other standards that can be used in layer 1 and layer 2 of S1 and X2 interfaces.

Figure 10.6 shows the Xn interface between the relay node and the donor eNodeB. Most of the protocols in these stacks are already familiar from earlier figures, but the physical layer is special in that it is a modified version of Uu, known as Un. The Un interface conforms to all normal physi-

10.6 E-UTRAN SPECIFICATIONS 213

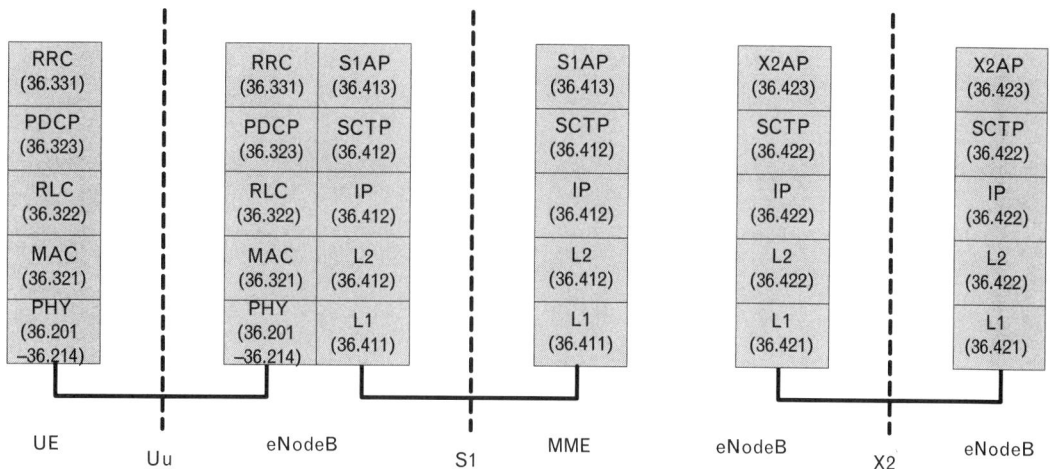

FIGURE 10.5 E-UTRAN interfaces, protocol tasks, and the corresponding specifications.

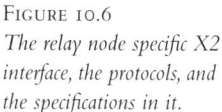

FIGURE 10.6
The relay node specific X2
interface, the protocols, and
the specifications in it.

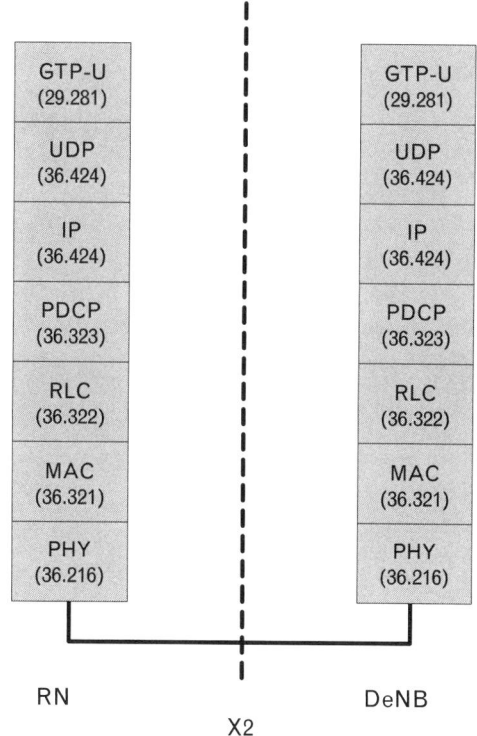

cal layer specifications: [2–5]. In addition, it implements the relay-specific definitions from [6].

The presentation in this section is limited to the most important protocol specifications. It lacks, for example, all radio performance and UE testing specifications.

The full list of LTE and LTE-A specifications in the 36 series is given in Appendix A.

References

[1] ITU-T Recommendation I.130 (1988): "Method for the Characterization of Telecommunication Services Supported by an ISDN and Network Capabilities of an ISDN."

[2] 3GPP TS 36.211, v 11.4.0, Evolved Universal Terrestrial Radio Access (E-UTRA); Physical Channels and Modulation, 09/2013.

[3] 3GPP TS 36.212, v 11.3.0, Evolved Universal Terrestrial Radio Access (E-UTRA); Multiplexing and Channel Coding, 06/2013.

[4] 3GPP TS 36.213, v 11.4.0, Evolved Universal Terrestrial Radio Access (E-UTRA); Physical Layer Procedures; 09/2013.

[5] 3GPP TS 36.214, v 11.1.0, Evolved Universal Terrestrial Radio Access (E-UTRA); Physical Layer; Measurements; 12/2012.

[6] 3GPP TS 36.216, v 11.0.0, Evolved Universal Terrestrial Radio Access (E-UTRA); Physical Layer for Relaying Operation; 09/2012.

Chapter 11

LTE-A Features

This chapter introduces a few selected features that have been adopted in LTE-A or are in the process of being adopted by LTE-A. Some of these features are not completely new in LTE-A (e.g., MIMO has been earlier specified for UMTS and LTE, too). However, these features have been given new capabilities in LTE-A, resulting in improved performance.

Also, this chapter includes some features that are completely new for 3GPP systems, such as coordinated multipoint (CoMP) transmission/reception and energy saving.

11.1 Energy Saving

This section discusses energy saving within the network infrastructure. Mobile devices are designed to be as energy saving as possible. They are typically wireless devices, running on batteries, and thus low power consumption is a very important design parameter. Long standby times are also exploited in marketing. Low UE power consumption is achieved by adding power saving features into system specifications and by clever hardware design.

On the network side the equipment is connected to mains power, and thus there is no similar operational requirement to save power as with mobile devices. However, electricity is getting increasingly expensive, and big mobile operators have a very large number of base stations in their networks. In addition, traffic patterns in mobile networks are very variable; peak demand takes place at different times in different locations. For example a shopping mall may have a high-capacity demand during prime shopping hours but close to zero demand at night time. Office blocks have high capacity demand during office hours, but low demand during weekends. Operators have to plan their networks so that they can meet the peak-time demand. However, this will leave them with unused capacity at other times.

As a consequence, mobile operators may want to switch off some of their base stations when they are not needed. Note that switching off a base station should not result in a coverage hole where mobile devices cannot get any service. Mobile networks, especially in capacity hotspots, have typically layered cell structure. The same area may be covered by both macro cells and capacity-boosting micro/pico/femto cells. If smaller cells are switched off, the macro cell(s) can still provide service for the remaining UEs (see Figure 11.1).

216   LTE-A FEATURES

FIGURE 11.1
Switching capacity cells on and off.

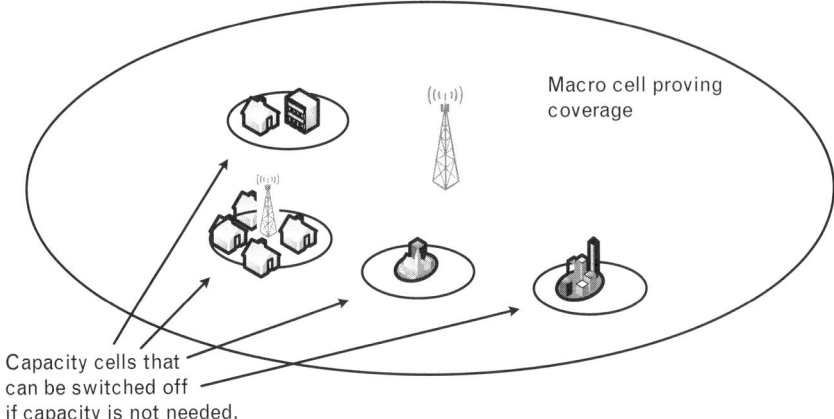

Mobile operators can, of course, switch their cells on and off already via their O&M systems. However, what has been missing so far are clearly defined mechanisms on how the operator can detect low capacity demand in a base station, how to switch it off, and most importantly when and how to switch it back on. This is not quite as simple as one would think. The principal problem is that mobile networks were not designed to have base stations that are continuously switched on and off. In addition, LTE does not have centralized control nodes in the radio access network, such as the RNC in UTRAN. The decision-making algorithm has to be distributed among all eNodeBs. They have to decide among themselves which eNodeB can be switched off and when, and what triggers will switch it back on. There are several problems to solve. For example:

1. Is it acceptable to switch off LTE cells, if the coverage is then provided by UMTS cells (or even EDGE cells)? This may result in some LTE-specific services becoming unavailable or to being provided with a lower quality of service. Note that there could also be LTE-only capable mobiles in the area, in which case those mobiles would be out of service.

2. Is a base station allowed to make the switch-off decision autonomously, or should this always be a centralized decision? (Note: base station switch-off is known as entering the dormant mode in 3GPP jargon.)

3. How/when should the base station make the switch-on decision? When the coverage base station detects increasing load, it has to switch on some capacity cells. However, capacity-boosting cells are typically much smaller than coverage-providing cells. How does the coverage cell know which capacity-boosting cells should be switched on to ease its loading? If it switches on the wrong cells with only a few or no UEs in their coverage area, the overload

11.1 ENERGY SAVING

situation in the coverage cell may continue. There are various methods to solve this problem, but it is not clear which ones would provide the best results. These methods include:

- OAM predefined policies;
- IoT measurements;
- UE measurements;
- Positioning information.

The first method is to let the O&M decide, based on operator-specific proprietary algorithms. This algorithm may simply involve two (or more) predefined sets of coverage configurations, store them in the O&M, and then reconfigure the network at set times using these configurations. For example, use configuration 1 at daytime and configuration 2 at night. The trigger to change the configuration may also be a certain loading pattern. Figure 11.2 gives an example of predefined configurations. In configuration 1, all capacity-boosting cells are switched off. This configuration is used at night. At 8:00 in the morning the cell(s) in the office district are switched on because extra capacity is needed when offices are opened. And in the afternoon around 3:00 the number of customers around the shopping district

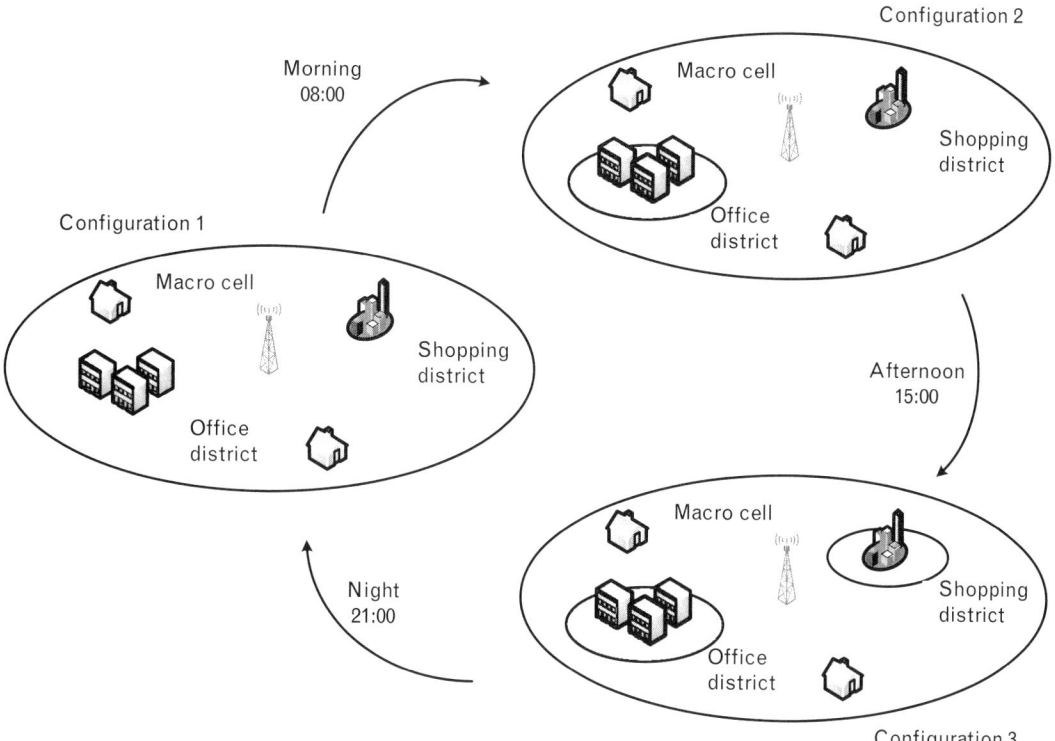

FIGURE 11.2   *Use of predefined configurations for energy saving.*

starts to increase and therefore the capacity cells are switched on also in this area. When the evening comes, both shops and offices will be closed, and thus all capacity cells can enter dormant mode at 9 pm.

Interference over thermal (IoT) measurements mean the coverage cell can ask a dormant cell or a set of dormant cells to switch on their listening capability and check the local noise level. If there is high level of noise, then it is probable that there are UEs in the area, and they would benefit if this cell is switched on again.

UE measurements method is where the coverage cell can ask a dormant cell or cells to transmit the reference signal for a short "probing" interval. At the same time, the UEs within the coverage cell are asked to measure the reference signal(s) and report back the results. After the probing interval, the dormant cells can return back to their dormant state. The measurement results should indicate to the coverage cell which dormant cells should be switched on.

The coverage cell can also exploit positioning information from UEs to determine which dormant cells to switch on. However, not all UEs have accurate positioning capability. To improve this method, dormant cells can start a specific timer when they are switched on and then on the expiration of this timer the cell should check if it is serving enough UEs to justify its switched-on state. If not, the cell can switch itself off again.

The eNodeB or eNodeBs that will provide the service once some other cells have been switched off to save energy are called compensating eNodeBs. Figure 11.2 presented a case where the compensating cell did not have to do anything to improve its coverage when the capacity cells were switched off since it already covered the area completely. Figure 11.3, on the other hand, presents a case where the compensating eNodeB will have to increase its Tx power to cover the whole coverage area of dormant cells. In this scenario, cells B, C, D, E, and F are switched off, and cell A has to increase its coverage area in compensation.

Once a decision has been made that the cell is to be switched off, all UEs with a connection in the cell have to be handed over to the compensation cell, and the energy-saving cell should not be switched off before this has been accomplished.

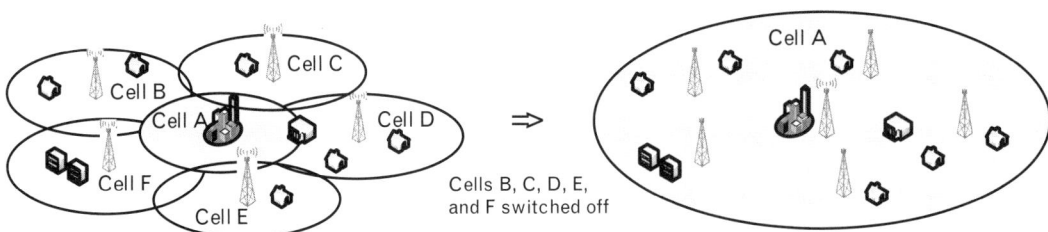

FIGURE 11.3 *Energy saving by increasing the cell size of the compensating cell.*

Note that base stations are the only part of the network infrastructure that can be switched off and on to provide energy saving. Other equipment in the network has to be switched on and in use continuously.

In summary, the problems in network energy saving include:

1. How to select which cells to switch off and when for energy saving;
2. How to decide which cells will act as compensating cells;
3. How to select which dormant cells to switch on and when.

The mechanisms to solve the issues above are still under discussion in Release 12. For example, this may require new signaling messages over the X2 interface between eNodeBs or it may be handled via O&M intervention, or the selected solution may involve both the O&M and the inter-eNodeB signaling.

Energy saving is a study item in Release 12, and, as such, standardized non-OAM solutions for network energy saving may have to wait until Release 13. Energy saving studies have so far produced two technical reports in [1, 2], and there is also a small section on energy saving in [3].

11.2 MIMO

11.2.1 MIMO Overview

MIMO is a form of multiantenna technology, exploiting a technique called spatial multiplexing. MIMO is not a 3GPP-specific technology and even within 3GPP, MIMO has already been used in UTRAN since Release 7 and in LTE from the beginning (i.e., Release 8). However, MIMO has been considerably enhanced for LTE-A, and that is why it is discussed here.

MIMO is not the only multiantenna technique, there are many other things one can do with multiple antennas. Sometimes the term MIMO has been misused, either deliberately or by ignorance, as a label for other multiantenna systems, too. In this book MIMO is considered to be a system that exploits spatial diversity (i.e., we do not consider MIMO to be synonymous to multiantenna systems that exploit beamforming or transmit and/or receiver diversity).

In spatial multiplexing each transmitting antenna sends independent data streams. MIMO introduces a new way of handling the radio interface channel resources. Previously transmission channels were thought to be shared and allocated among users by partitioning of frequency and time (as in FDMA/TDMA systems such as GSM) or by means of frequency, time, and code (as in UMTS). However, MIMO introduces a new spatial dimension. It has been shown that it is possible to separate two or more transmissions at the receiver even if they have been sent using the same frequency, time, and code, as long as their spatial signatures are sufficiently different. MIMO systems can achieve this by using several transmit and receive antennas, provided that the antenna spacing at both ends of the radio link is sufficiently large and the environment is rich in scatterers. In optimal conditions these antennas can form several parallel transmission channels, which can employ the same frequency, time, and code space, thus increasing the system capacity considerably. A MIMO system with m transmit and n receive antennas can achieve up to C = min (m,n) independent subchannels. In terms of MIMO, a good operating environment is one with high signal-to-noise (S/N) ratio and that is rich in multipath (for an example, see Figure 11.4).

However, increasing the number of independent subchannels by means of increasing the number of transmit and receive antennas will bring diminishing returns. The channels have to be uncorrelated and uncoupled, both of which are increasingly difficult to achieve when more antennas needs to be packed into small devices. The bigger the distance between antennas, the more likely it is that the transmissions from those antennas will undergo independent fading and thus they will have unique spatial signatures. MIMO antenna technology has developed rapidly in recent years, and there are several different types of antennas available, each with different performance parameters. However, a good ballpark figure for UE design is that antenna elements should be at least half a wavelength apart. For a mobile operating at 2.6-GHz frequency, this means about 5.8 cm. But on 800-MHz operating band the separation should already be 18.7 cm. Lower frequencies are clearly challenging for MIMO handsets. But it is also good to remember that data-hungry devices are not those with the smallest form factor, but those with larger displays: large smart phones and tablet PCs. Such devices provide a much better base onto which to attach MIMO antennas. Still, multiple MIMO antenna elements are problematic, and thus further research on multiantenna techniques is needed.

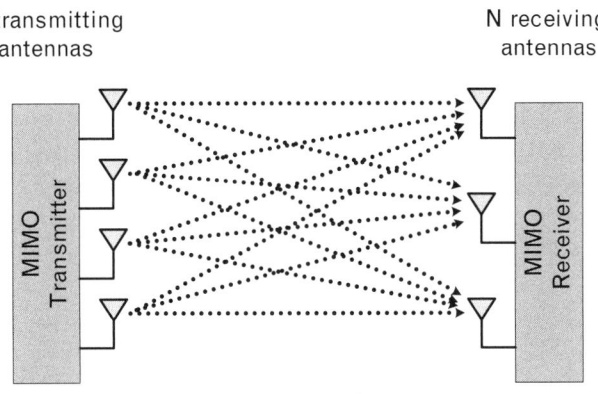

FIGURE 11.4
General MIMO system picture.

11.2.2 Downlink MIMO

LTE in Releases 8 and 9 supported 4 × 4 MIMO in the downlink and 2 × 2 MIMO in the uplink. In LTE-A this is increased to 8 × 8 in the downlink and 4 × 4 in the uplink. The 8 × 8 MIMO is definitely something that will not be seen in operational networks in the near future, since equipping both eNBs and UEs with eight antennas is a costly and complex exercise.

The theoretical peak data rate for an 8 × 8 MIMO system deployed on a 20-MHz carrier would be 600 Mbps. LTE-A also enables carrier aggregation, up to five component carriers together. If we assume an 8 × 8 MIMO, and 5 × 20 MHz aggregated carriers, then the theoretical maximum is 3 Gbps. Needless to say, this is a purely theoretical number.

There are two ways to use MIMO in 3GPP: single user MIMO (SU-MIMO) and multiuser MIMO (MU-MIMO) (see Figure 11.5). In SU-MIMO all MIMO layers are assigned to the same UE; in MU-MIMO they are assigned to different UEs. A cell can switch between these modes dynamically. The UE in a MU-MIMO cell does not have to be aware of other UEs, which may be spatially multiplexed to use the same frequency/time resources.

SU-MIMO provides a higher theoretical peak data rate for one user, but in practice the full 8 × 8 SU-MIMO will not be used for some time to come. Its successful usage would require that those several layers are not spatially correlated and indeed that both the eNB and the UE have 8 Tx/Rx antennas. Moreover, most applications do not require or provide continuous high data rates. Therefore using SU-MIMO with an 8 × 8 antenna installation (or even with 4 × 4) is likely to result in a suboptimal solution system capacity-wise. MU-MIMO is a better solution in this sense. MU-MIMO can be exploited even by single-antenna UEs; that means it is available for all LTE-A-capable UEs. Moreover, since the individual UEs are in different locations, it is more likely that the different MIMO layers are spatially

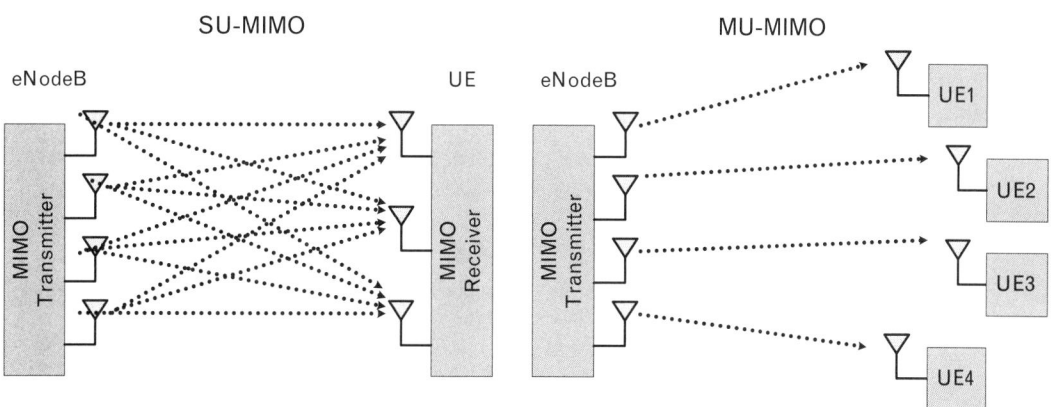

FIGURE 11.5 *SU-MIMO and MU-MIMO.*

separated. As a result MU-MIMO is likely to provide higher throughput per cell/sector than SU-MIMO.

Spatial multiplexing in LTE downlink can be either open loop or closed loop. Open loop spatial multiplexing does not need channel state information (measurement results) from the UE, whereas in the closed loop method UE feedback is required. Generally speaking, if the transmitter knows the channel state information (CSI), it can use this information in precoding, which will result in more robust transmissions. The transmitter tries to compensate for the problems in the airwaves in such a way that the signal that the receiver gets should be as error-free as possible.

There are two major problems in delivering the CSI from the receiver to the transmitter:

1. There are many transmission channels, especially in MIMO systems, and thus several independent channels to measure. The amount of measurement results very easily grows so large that its delivery to the transmitter would consume a large amount of available bandwidth.
2. The channel state can change very quickly, especially if the receiver or transmitter is moving. Therefore, channel measurements should be delivered to the transmitter quickly because they will become less and less relevant over time.

Problem 1 is an especially serious one. A 2×2 MIMO system has 4 individual component channels to measure and report, a 4×4 system has 16 channels, and a 8×8 MIMO system should report the channel state information from 64 channels. In LTE the problem of CSI delivery has been solved by the use of codebooks. A codebook is a collection of predefined matrices, each indicating a certain set of the transmission parameters that are suitable for this channel state. The receiver measures the channel and then selects the entry from the codebook that best characterizes the CSI. Instead of signaling all the channel parameters to the transmitter, the receiver simply sends the index of the codebook entry. A codebook has a limited size, and therefore this method can never fully characterize the channel. In a 2×2 MIMO channel the codebook size is 4, and in a 4×4 system it is 16. Thus, a 2×2 MIMO receiver can signal the codebook index with two bits, and a 4×4 MIMO receiver needs four bits for the index. The codebooks are defined in [4].

Once the precoder in the transmitter knows the precoding matrix index to be used, it simply takes the block of symbol vectors from the layer mapping as an input and multiplies it with the precoding matrix.

Note that in the TDD mode the problem of CSI is much easier to solve since the same channel is used for both transmission links (uplink and downlink). The transmitter will know the channel state because it is also receiving on the same channel.

The use of multiple antennas is under the control of the eNodeB. In order to be able to quickly inform the UE on what kind of data transmission method is used in the downlink, a concept called the *transmission mode* (TM) has been defined. In Release 11 there are 10 different TMs. The eNodeB informs the UE about the TM to be used via RRC signaling. Different TMs indicate whether multiple antennas are used, and, if they are used, then in what kind of configuration (e.g., transmit diversity or spatial multiplexing multiuser MIMO), and other precoding-related information. Table 11.1 [5] gives a short description of each mode.

In general, MIMO (or spatial multiplexing in standards jargon) should be used when the channel conditions are good, and transmit diversity or large delay cyclic delay diversity (CDD) should be used when the channel conditions are poor. This is because spatial multiplexing requires a high-quality radio channel to perform successfully. On weaker channels, it is better to transmit only one data stream and use diversity provided by multiple antennas to make sure that the data gets through without errors. On weak channels diversity schemes can provide higher data throughput than spatial multiplexing because it may be possible to employ more efficient MCS val-

TABLE 11.1 TRANSMISSION MODES IN THE DOWNLINK

TRANSMISSION MODE	TRANSMISSION SCHEME OF PDSCH
1	Single-antenna port, port 0
2	Transmit diversity
3	Transmit diversity if the associated rank indicator is 1; otherwise, large delay CDD
4	Closed-loop spatial multiplexing
5	Multiuser MIMO
6	Closed-loop spatial multiplexing with a single transmission layer
7	If the number of PBCH antenna ports is one single-antenna port, port 0; otherwise, transmit diversity
8	If the UE is configured without PMI/RI reporting: if the number of PBCH antenna ports is one single-antenna port, port 0; otherwise, transmit diversity
	If the UE is configured with PMI/RI reporting: closed-loop spatial multiplexing
9	If the UE is configured without PMI/RI reporting: if the number of PBCH antenna ports is one single-antenna port, port 0; otherwise, transmit diversity
	If the UE is configured with PMI/RI reporting: if the number of CSI-RS ports is one single-antenna port, port 7; otherwise, up to eight-layer transmission, ports 7–14 (see subclause 7.1.5B)
10	If a CSI process of the UE is configured without PMI/RI reporting: if the number of CSI-RS ports is one single-antenna port, port 7; otherwise, transmit diversity
	If a CSI process of the UE is configured with PMI/RI reporting: if the number of CSI-RS ports is one single-antenna port, port 7; otherwise, up to eight-layer transmission, ports 7–14 (see subclause 7.1.5B)

ues (e.g., higher-order modulation) with diversity schemes than with spatial multiplexing (Figure 11.6).

11.2.3 Uplink MIMO

Uplink MIMO in LTE-A is slightly different from downlink MIMO. First of all, UEs are physically much smaller than eNodeBs and thus it is difficult to have as many antennas in a UE as in an eNodeB. Therefore, the maximum number of antennas is four, at least in Release 11. Moreover, UEs are handheld devices, and thus it is impossible to employ techniques such as beamforming in the uplink. Therefore, in the uplink there are just two alternatives on what the UE can do with its antennas: single antenna transmission or spatial multiplexing with two to four antennas.

MIMO is discussed in several LTE specifications (e.g., [3–6]).

11.3 Relays

Relays in LTE-A are so called L3 relays. These are quite intelligent devices, combining both eNB and UE functionalities. They are different from much simpler repeaters, also known as L1 relays, which basically just forward the received signal blindly.

The reasons to use relays in LTE-A are that they can provide enhanced coverage and capacity at cell edges and they can also be used to connect remote locations without fiber connections. The latter point is actually very important in some cases. Small base stations as such are quite inexpensive nowadays, but the backhaul connection they require can, in some cases, be very expensive to install. Relay nodes (RNs) have the wireless backhaul connection "in-built" in the technology, though a 3GPP RN is not by all

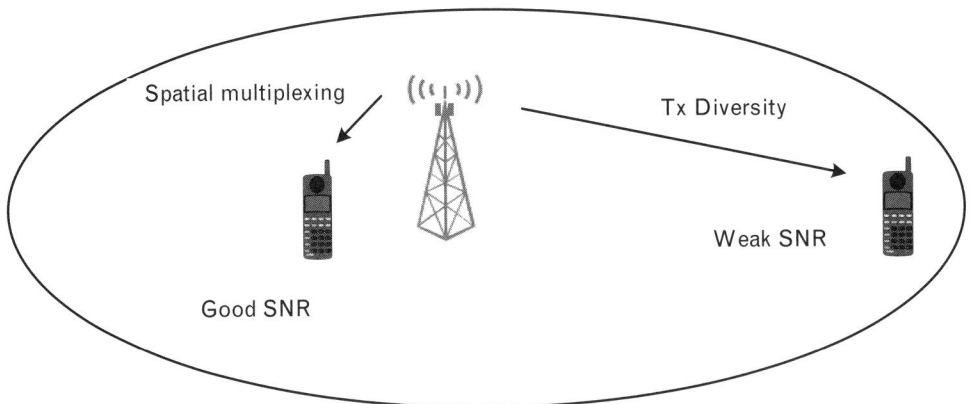

FIGURE 11.6 *Spatial multiplexing versus transmit diversity.*

11.3 RELAYS

means the only technology that can be used to build wireless backhaul for base stations.

Figure 11.7 presents the LTE-A relay architecture. The relay node is connected to the donor eNB (DeNB) via the radio interface Un. This is a modified LTE air interface Uu. The difference between Uu and Un is specified in [7]. From the UE point of view, the RN behaves like an eNB. In fact, the UE is not aware that it is connected to a RN; this entity is just another eNB for a UE. The RN supports the eNB functionality as well as part of the UE functionality. The RN terminates S1, X2, and Un interfaces. The DeNB provides S1 and X2 proxy functionality between the RN and other network nodes such as eNBs, MMEs, and S-GWs. Because of the proxy functionality, the DeNB appears as an MME (for S1-MME), an eNB (for X2), and an S-GW (for S1-U) to the RN. The protocol stacks in a RN were presented in Section 7.4.

At the moment (Release 12) LTE-A does not support multihop relays (i.e., setups where several RNs are linked into a chain). There is only one RN between the DeNB and the UE. Also, RN mobility is not supported in Releases 10 and 11 (that is, an RN cannot change its DeNB while it is in active mode), although this may change in the future with mobile relays (see Section 12.2.3).

With relay nodes, there are two radio links: Un interface for the backhaul link and Uu interface for the access link. These can operate on the same frequency or on different frequencies. Both arrangements have their advantages and disadvantages.

Same-frequency relays are called inband relays. These are more spectrum efficient, since both the DeNB and the RN use the same frequency band. Therefore, this is a good solution for cases where spectrum is scarce. On the other hand, inband relays need to employ a special subframe configuration

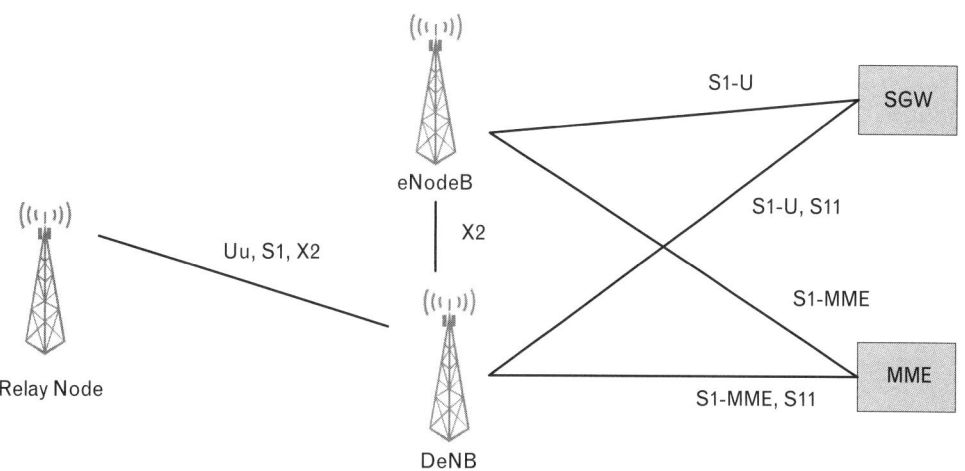

FIGURE 11.7 *LTE relay architecture.*

that separates backhaul and access links in the time domain. The inband solution also causes longer delays to signals because of the necessary time delay between backhaul and access signals.

If the backhaul and access links are configured to use different frequencies, then the result is called an outband relay. In this case, subframe configuration data is needed. Both radio links can operate independently of each other, though the penalty is increased spectrum usage. Also care should be taken that the two frequencies are far enough apart from each other to prevent interference (see Figure 11.8).

The third option is to employ inband relays, but to isolate the antennas for backhaul and access links in a way that prevents interantenna interference (e.g., by using directed antennas). In this case, a special subframe configuration is not needed.

Which relay configuration is used (inband or outband) will be configured by the O&M system. Once the RN is set up, it will inform the DeNB in the RRC connection setup complete message (rn-SubframeConfigReq field) whether it needs to receive subframe configuration data.

Relay functionality is specified in [3, 7–10].

11.4 Carrier Aggregation

Carrier aggregation is a new feature that is adopted in LTE-advanced. It is used to combine several carriers (up to five) into one logical aggregated carrier. However, in Releases 10 and 11 only two component carriers are supported. The component carriers are normal carriers (i.e., similar to standard LTE carriers), and thus they can have a bandwidth of 1.4, 3, 5, 10, 15, or 20 MHz. Therefore, the theoretical maximum bandwidth of an aggregated

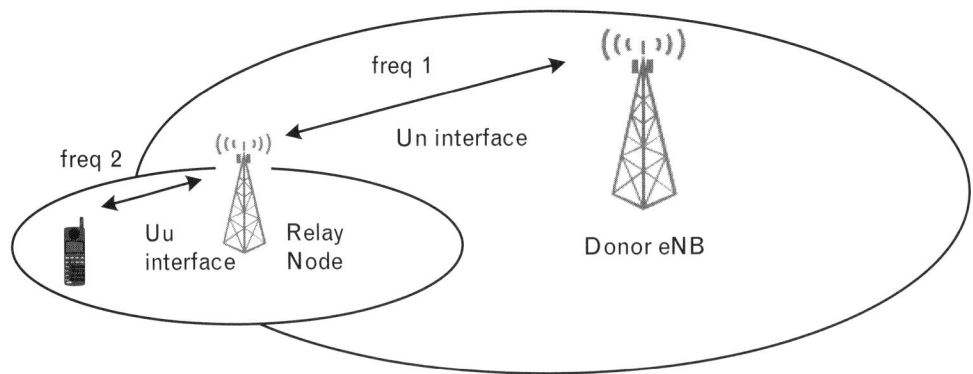

FIGURE 11.8 *Inband and outband relays.*

carrier is 100 MHz. However, it is very unlikely that such a wide carrier is seen in operational networks. Rather, the main reason for carrier aggregation is that many operators have access to only relatively small slices of spectrum, and carrier aggregation provides to way to combine those slices into wider spectrum channels.

The component carriers do not have to be similar in size. For example, one aggregated carrier can consist of three component carriers with bandwidths of 3, 5, and 10 MHz. In the FDD mode the number of component carriers can be different in uplink and downlink, but in the uplink the number of component carriers has to be smaller than or equal to the number of downlink component carriers. In the TDD mode, due to the nature of this mode, both uplink and downlink must have the same number of components carriers.

There are three ways to combine component carriers into an aggregated carrier:

1. Intraband contiguous;
2. Intraband noncontiguous;
3. Interband noncontiguous.

In the first case all component carriers are on the same frequency band, forming a continuous block. In the second case the component carriers are still on the same frequency band, but there are frequency gaps between them. In the third case the component carriers are from different frequency bands, in which case there are obviously gaps between them (see Figure 11.9).

Since components carriers are standard LTE carriers, older LTE (non-LTE-A) devices (i.e., release 8/9 UEs) can use them as before, whereas LTE-A devices can exploit also aggregate carriers.

Each component carrier employs a separate serving cell. However, these cells may and often will originate from the same eNodeB. One of the serving cells is the primary component carrier (PCC), and the others are secondary component carriers (SCC). The RRC connection (control signaling) is handled by the PCC only.

Component carriers may have different coverage areas. This is obvious if they originate from different eNodeBs, but even CCs originating from the same eNodeB can differ in their coverage areas. This is due to different carrier frequencies and different transmission powers. Release 11 introduced different timing advance values for different component carriers, which enables the CCs to originate from multiple sites. Furthermore, this feature enables the use of CCs in heterogenous networks (HetNets); (i.e., networks that have a mixture of macro, pico, and femto cells). Traditionally such networks are implemented in a way that each layer (macro, pico, femto) has its own frequency band in order to avoid interference between

228 LTE-A FEATURES

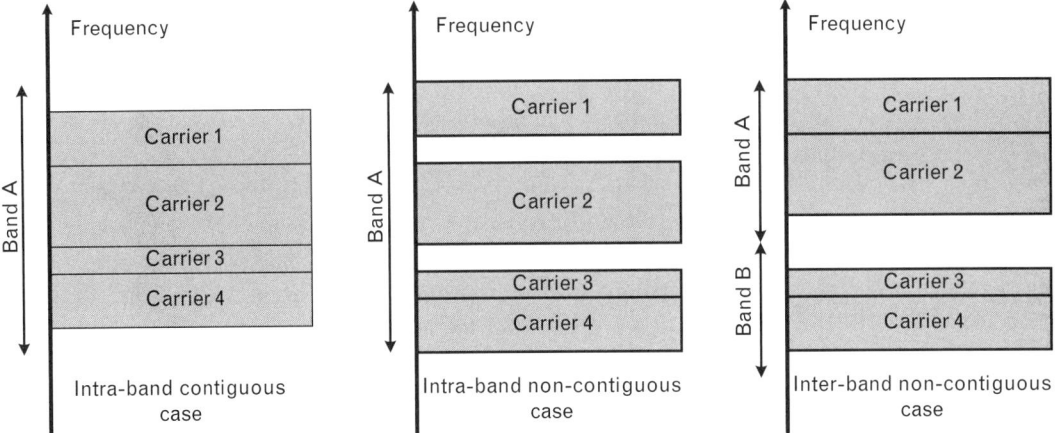

FIGURE 11.9 *Intra- and interband carrier aggregation.*

layers. However, this may lead to less than optimal resource usage and also numerous handovers when a UE moves between layers. However, carrier aggregation enables the UE to have CCs on different layers. Typically, the macro cell is the PCC, and pico/femto cells are SCCs. This arrangement enables the UE to receive control signaling via the macro cell as well as user data via small cells that probably have more capacity available (see Figure 11.10).

Since in carrier aggregation the PCC will handle control signaling for all CCs, it is foreseen that the capacity of the Release 8 PDCCH may run out (there are also other reasons why the PDCCH may become highly loaded, such as MU-MIMO). To alleviate the problem, Release 11 has also introduced a new control channel type, enhanced physical downlink control channel (ePDCCH). The extra capacity for this channel is taken from PDSCH resources.

The allowed component carrier combinations for carrier aggregation are strictly defined; various carriers cannot be combined at will. This is because each combination has different radio performance requirements. Carrier aggregation is designed to operate in operating bands defined in

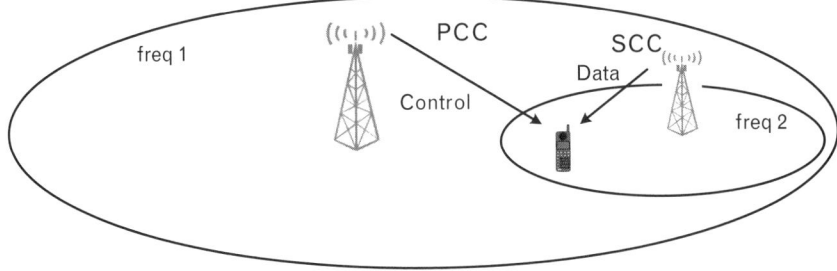

FIGURE 11.10 *Carrier aggregation in a multisite HetNet.*

11.5 Enhanced Intercell Interference Coordination

Tables 11.2, 11.3, and 11.4 [11]. But note that these tables are continuously evolving as new carrier combinations are added to them.

There are six bandwidth classes defined for carrier aggregation, though only three of them are in use so far (A, B, and C). As seen from Table 11.5, the maximum number of component carriers is only two in classes B and C.

Carrier aggregation as a feature is much more than just combining individual component carriers into bigger channels. Its ability to make HetNets to operate more efficiently means that carrier aggregation will be widely used in future LTE networks. Carrier aggregation is discussed in [3, 11, 12, 13].

11.5 Enhanced Intercell Interference Coordination

Enhanced intercell interference coordination (eICIC) tries to mitigate intercell interference at cell edges. The ICIC feature actually consists of three (so far) different phases. The original ICIC was introduced in Release 8, the enhanced version (eICIC) in Release 10, and the further enhanced version (FeICIC) in Release 11.

The original ICIC is about scheduling the transmission resources (resource blocks and transmission power) in a way that mitigates the intercell interference. The basic idea is not to allocate the same resource blocks to UEs on both sides of a cell boundary. Instead it is better to perform intercell resource allocation coordination for users near cell edges. Users that are closer to the eNodeB do not need similar treatment because they are less likely to cause intercell interference, as they are further away from cell boundaries, and in addition their transmission power is likely to be low because they are close to the eNodeB. Figure 11.11 clarifies a typical ICIC solution. Cell center UEs can be allocated any resource blocks and low tx power. On the other hand, the UEs near cell edges must use higher tx power to reach the eNodeB, and they can only be allocated a subset of the available resource blocks—a subset that does not overlap with the resource

Table 11.2 Intraband Contiguous CA Operating Bands

E-UTRA CA Band	E-UTRA Band	Uplink (UL) Operating Band BS Receive/UE Transmit $F_{UL\_low} - F_{UL\_high}$			Downlink (DL) Operating Band BS Transmit/UE Receive $F_{DL\_low} - F_{DL\_high}$			Duplex Mode
CA_1	1	1920 MHz	–	1980 MHz	2110 MHz	–	2170 MHz	FDD
CA_3	3	1710 MHz	–	1785 MHz	1805 MHz	–	1880 MHz	FDD
CA_7	7	2500 MHz	–	2570 MHz	2620 MHz	–	2690 MHz	FDD
CA_38	38	2570 MHz	–	2620 MHz	2570 MHz	–	2620 MHz	TDD
CA_40	40	2300 MHz	–	2400 MHz	2300 MHz	–	2400 MHz	TDD
CA_41	41	2496 MHz	–	2690 MHz	2496 MHz	–	2690 MHz	TDD

TABLE 11.3 INTERBAND CA OPERATING BANDS

E-UTRA CA Band	E-UTRA Band	Uplink (UL) Operating Band BS Receive/UE Transmit $F_{UL\_low} - F_{UL\_high}$	Downlink (DL) Operating Band BS Transmit/UE Receive $F_{DL\_low} - F_{DL\_high}$	Duplex Mode
CA_1-5	1	1920 MHz – 1980 MHz	2110 MHz – 2170 MHz	FDD
	5	824 MHz – 849 MHz	869 MHz – 894 MHz	
CA_1-8	1	1920 MHz – 1980 MHz	2110 MHz – 2170 MHz	FDD
	8	880 MHz – 915 MHz	925 MHz – 960 MHz	
CA_1-18	1	1920 MHz – 1980 MHz	2110 MHz – 2170 MHz	FDD
	18	815 MHz – 830 MHz	860 MHz – 875 MHz	
CA_1-19	1	1920 MHz – 1980 MHz	2110 MHz – 2170 MHz	FDD
	19	830 MHz – 845 MHz	875 MHz – 890 MHz	
CA_1-21	1	1920 MHz – 1980 MHz	2110 MHz – 2170 MHz	FDD
	21	1447.9 MHz – 1462.9 MHz	1495.9 MHz – 1510.9 MHz	
CA_1-26	1	1920 MHz – 1980 MHz	2110 MHz – 2170 MHz	FDD
	26	814 MHz – 849 MHz	859 MHz – 894 MHz	
CA_2-4	2	1850 MHz – 1910 MHz	1930 MHz – 1990 MHz	FDD
	4	1710 MHz – 1755 MHz	2110 MHz – 2155 MHz	
CA_2-5	2	1850 MHz – 1910 MHz	1930 MHz – 1990 MHz	FDD
	5	824 MHz – 849 MHz	869 MHz – 894 MHz	
CA_2-13	2	1850 MHz – 1910 MHz	1930 MHz – 1990 MHz	FDD
	13	777 MHz – 787 MHz	746 MHz – 756 MHz	
CA_2-17	2	1850 MHz – 1910 MHz	1930 MHz – 1990 MHz	FDD
	17	704 MHz – 716 MHz	734 MHz – 746 MHz	
CA_2-29	2	1850 MHz – 1910 MHz	1930 MHz – 1990 MHz	FDD
	29	N/A	717 MHz – 728 MHz	
CA_3-5	3	1710 MHz – 1785 MHz	1805 MHz – 1880 MHz	FDD
	5	824 MHz – 849 MHz	869 MHz – 894 MHz	
CA_3-7	3	1710 MHz – 1785 MHz	1805 MHz – 1880 MHz	FDD
	7	2500 MHz – 2570 MHz	2620 MHz – 2690 MHz	
CA_3-8	3	1710 MHz – 1785 MHz	1805 MHz – 1880 MHz	FDD
	8	880 MHz – 915 MHz	925 MHz – 960 MHz	
CA_3-19	3	1710 MHz – 1785 MHz	1805 MHz – 1880 MHz	FDD
	19	830 MHz – 845 MHz	875 MHz – 890 MHz	
CA_3-20	3	1710 MHz – 1785 MHz	1805 MHz – 1880 MHz	FDD
	20	832 MHz – 862 MHz	791 MHz – 821 MHz	
CA_3-26	3	1710 MHz – 1785 MHz	1805 MHz – 1880 MHz	FDD
	26	814 MHz – 849 MHz	859 MHz – 894 MHz	
CA_3-28	3	1710 MHz – 1785 MHz	1805 MHz – 1880 MHz	FDD
	28	703 MHz – 748 MHz	758 MHz – 803 MHz	
CA_4-5	4	1710 MHz – 1755 MHz	2110 MHz – 2155 MHz	FDD
	5	824 MHz – 849 MHz	869 MHz – 894 MHz	
CA_4-7	4	1710 MHz – 1755 MHz	2110 MHz – 2155 MHz	FDD
	7	2500 MHz – 2570 MHz	2620 MHz – 2690 MHz	

11.5 ENHANCED INTERCELL INTERFERENCE COORDINATION

TABLE 11.3 (CONTINUED)

E-UTRA CA Band	E-UTRA Band	Uplink (UL) Operating Band BS Receive/UE Transmit $F_{UL\_low} - F_{UL\_high}$	Downlink (DL) Operating Band BS Transmit/UE Receive $F_{DL\_low} - F_{DL\_high}$	Duplex Mode
CA_4-12	4	1710 MHz – 1755 MHz	2110 MHz – 2155 MHz	FDD
	12	699 MHz – 716 MHz	729 MHz – 746 MHz	
CA_4-13	4	1710 MHz – 1755 MHz	2110 MHz – 2155 MHz	FDD
	13	777 MHz – 787 MHz	746 MHz – 756 MHz	
CA_4-29	4	1710 MHz – 1755 MHz	2110 MHz – 2155 MHz	FDD
	29	N/A	717 MHz – 728 MHz	
CA_5-12	5	824 MHz – 849 MHz	869 MHz – 894 MHz	FDD
	12	699 MHz – 716 MHz	729 MHz – 746 MHz	
CA_5-17	5	824 MHz – 849 MHz	869 MHz – 894 MHz	FDD
	17	704 MHz – 716 MHz	734 MHz – 746 MHz	
CA_7-20	7	2500 MHz – 2570 MHz	2620 MHz – 2690 MHz	FDD
	20	832 MHz – 862 MHz	791 MHz – 821 MHz	
CA_8-20	8	880 MHz – 915 MHz	925 MHz – 960 MHz	FDD
	20	832 MHz – 862 MHz	791 MHz – 821 MHz	
CA_11-18	11	1427.9 MHz – 1447.9 MHz	1475.9 MHz – 1495.9 MHz	FDD
	18	815 MHz – 830 MHz	860 MHz – 875 MHz	
CA_19-21	19	830 MHz – 845 MHz	875 MHz – 890 MHz	FDD
	21	1447.9 MHz – 1462.9 MHz	1495.9 MHz – 1510.9 MHz	
CA_23-29	23	2000 MHz – 2020 MHz	2180 MHz – 2200 MHz	FDD
	29	N/A	717 MHz – 728 MHz	

TABLE 11.4 INTRABAND NONCONTIGUOUS CA OPERATING BANDS

E-UTRA CA Band	E-UTRA Band	Uplink (UL) Operating Band BS Receive/UE Transmit $F_{UL\_low} - F_{UL\_high}$	Downlink (DL) Operating Band BS Transmit/UE Receive $F_{DL\_low} - F_{DL\_high}$	Duplex Mode
CA_4-4	4	1710 MHz – 1755 MHz	2110 MHz – 2155 MHz	FDD
CA_25-25	25	1850 MHz – 1915 MHz	1930 MHz – 1995 MHz	FDD
CA_41-41	41	2496 MHz – 2690 MHz	2496 MHz – 2690 MHz	TDD

block subset of the neighboring cell. In this example, UE2 can be allocated any resource blocks, whereas for UE1 only those resource blocks that belong to cell B specific subset can be allocated.

The system needs a minimum of three different resource block subsets because the subsets can be reused in nonneighboring cells.

ICIC works fine for standard macrocell deployments, but not in small pico/femto cells that are deployed inside macrocells. As was shown in Section 11.4, carrier aggregation could provide a solution to HetNet intercell

232 LTE-A FEATURES

TABLE 11.5 Carrier Aggregation Bandwidth Classes

CA Bandwidth Class	Aggregated Transmission Bandwidth Configuration	Maximum number of CC
A	$N_{RB,agg} \leq 100$	1
B	$N_{RB,agg} \leq 100$	2
C	$100 < N_{RB,agg} \leq 200$	2
D	$200 < N_{RB,agg} \leq [300]$	FFS
E	$[300] < N_{RB,agg} \leq [400]$	FFS
F	$[400] < N_{RB,agg} \leq [500]$	FFS

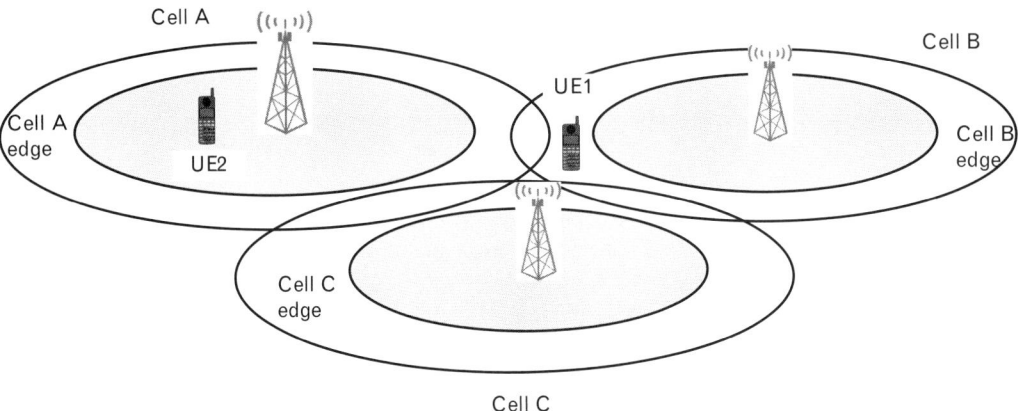

FIGURE 11.11 *ICIC resource allocation.*

interference problems, but if carrier aggregation is not used, then eICIC can be employed.

Due to the macro cell's large transmit power, the effective range of the femto cell will be reduced as shown in Figure 11.12. This can be reversed by assigning a bias in the UE's cell selection algorithm. This bias can be

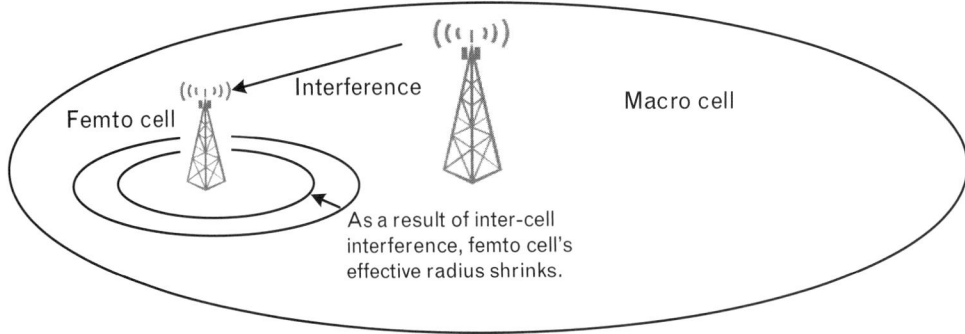

FIGURE 11.12 *Heterogeneous network intercell interference problem.*

up to 6 dB in Release 10. Though it can help the UEs at femto cell edges reconnect to the femto cell, it does not help with the interference those UEs are experiencing. The solution proposed in Release 10 is to use almost blank subframes (ABS). A certain number of subframes are nominated as ABS subframes. During ABS subframes only the most minimal set of signals (CRS/PSS/SSS/PBCH) is transmitted from the macro cell. The femto cell can then use these quiet frames to transmit to the UEs in its coverage area. The ABS configuration can be communicated to the femtocell via the X2 interface and via the OAM interface if the X2 interface is not available.

Note that the ABS approach applies only to UEs that are in the femto cell edge area. UEs that are near the cell center can be served also during non-ABS subframes.

However, in practisc the heterogeneous network intercell interference problem is not this serious. Namely, femtocells are often deployed indoors, and then the interference from macrocells is not that severe. The walls of the building may act as a natural cell boundary; indoors are served by the femtocell and outdoors by the macrocell.

In Release 11 the intercell interference cancellation was further enhanced in the FeICIC work item. The main enhancement was to provide the UE with cell specific reference symbol (CRS) assistance information of the interfering cells. This information can be used in the interference cancellation receiver in the UE.

In case of strong interference the UE may not be able to decode even the system information transmitted. Therefore in Release 11 it is possible to transmit system information block type 1 (SIB1) information via dedicated RRC signaling (RRCConnectionReconfiguration message). SIB1 includes important information like PLMN IDs, tracking area code, cell identity, access restrictions, and information on scheduling of all other system information elements.

For further information, please check [3, 14].

11.6 Evolved Multimedia Broadcast Multicast Service

Evolved multimedia broadcast multicast service (eMBMS) was first introduced to LTE in Release 9. MBMS never took off in UMTS because UMTS networks simply did not have enough capacity to support adequate MBMS services. But with greatly enhanced capabilities of LTE-A, the capacity should not become an obstacle with eMBMS.

The most important eMBMS enhancement in Release 10 is the counting function. This feature is designed to help the network operator to decide which MBMS services are of interest to users and to prioritize them accordingly. Because MBMS is a broadcast/multicast service, the network does not necessarily know how many users (if any) are receiving a particular broadcast. This may lead to a waste of resources if the network broadcasts services nobody (or very few users) wants to receive. The counting function checks the number of connected mode users that either already receive or want to receive a particular MBMS service.

The counting procedure is initiated by the network. The multicell/multicast coordination entity (MCE) requests the eNodeBs with an MBMS service counting request message to count and report for one or more MBMS services the number of connected mode UEs receiving the MBMS service or interested in receiving it.

The connected mode UEs that are interested in the indicated services will respond with a RRC counting response message. This message contains short MBMS service identities (unique within the MBSFN service area) and may optionally include the information to identify the MBSFN area (if overlapping is configured).

Only Release 10 and later UEs are counted; Release 9 UEs will be ignored, since they do not support the necessary procedures. Also, only connected mode UEs are counted.

The results are delivered back to the MCE in an MBMS service counting results reporting message. The MCE then decides the priority order of different MBMS services and allocates channel resources accordingly.

MBMS service, related interfaces, and related procotols are discussed in [3, 15–20]. See Figure 11.13 for a top-level MBMS architecture.

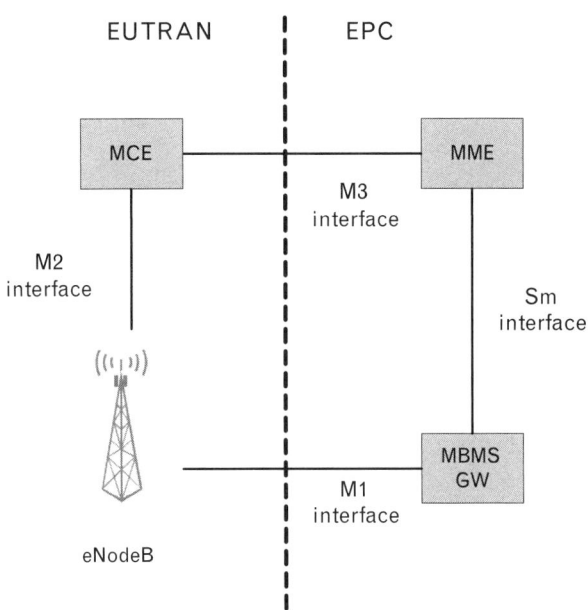

FIGURE 11.13
High-level MBMS architecture.

11.7 Self-Organizing Networks

SON is a set of various automatic tools that the operator can use to ease the deployment and management of LTE networks. These tools are very varying in nature, and some have very little in common. Yet they all belong under SON umbrella.

Next, we discuss the most important SON improvements in Releases 10 and 11.

11.7.1 Automatic Neighbour Relations (ANR)

The purpose of the ANR function is to relieve the operator from the task of manually managing neighbor relations (NRs) of cells. The ANR function is located in the eNodeB and manages the neighbor relation table (NRT). With the help of UEs, the neighbor detection function finds new neighbors and adds them to the NRT. ANR also contains the neighbor removal function, which removes outdated NRs.

Release 11 ANR improvements include interradio access technology (IRAT) ANR (i.e., ANR from E-UTRAN to GERAN, UTRAN, and CDMA2000).

11.7.2 Mobililty Load Balancing (MLB)

The objective of load balancing is to distribute cell load evenly among cells or to transfer part of the traffic from congested cells, at the same time avoiding unnecessary load balancing handovers and redirections. This is done by self-optimizing mobility parameters and handover actions. The MLB function includes the following subfunctions:

- Load reporting;
- Load balancing handovers;
- Load balancing by adapting handover and/or reselection parameters.

11.7.3 Mobility Robustness Optimization (MRO)

The MRO function aims at detecting and enabling correction of the following problems:

- Connection failures due to intra-LTE or inter-RAT mobility;
- Unnecessary handover (HO) to another RAT (too early IRAT HO with no radio link failure);

- Inter-RAT ping-pong HO.

This function aims at to optimize relevant HO algorithm parameters to mitigate the problems of too early handovers, too late handovers, and inefficient use of network resources.

11.7.4 Coverage and Capacity Optimization

This function identifies problems with network coverage or capacity. The network may have coverage holes, weak coverage, pilot pollution, overshoot coverage, or DL and UL channel coverage mismatch. The input data for the function is collected from performance measurements at the source and/or target eNodeBs, minimizing drive tests (MDT) measurements, or handover (HO)–related performance measurements.

11.7.5 RACH Optimization Function

This function aims to automatically set several parameters related to the performance of random access channel (RACH):

- RACH configuration (resource unit allocation);
- RACH preamble split (among dedicated, group A, group B);
- RACH backoff parameter value;
- RACH transmission power control parameters.

11.7.6 Coordination Between Various SON Functions

This function addresses conflicts that could happen when separate SON functions try to change the same network configuration parameter. This is a new SON function, specified in Release 11.

Energy saving and MDT functions are also sometimes included in SON functions. Both of these have been discussed elsewhere in this book; energy saving in Section 11.1 and MDT in Section 6.5.3.7.

SON is discussed in several specifications because SON consists of many independent functions. For a start, see [3, 21].

11.8 Coordinated Multipoint Transmission/Reception

Coordinated multipoint transmission and reception function (CoMP) resembles the soft handover function in some other mobile networks. It potentially improves the quality of service for UEs that are near cell edges.

11.8 COORDINATED MULTIPOINT TRANSMISSION/RECEPTION

Such UEs may be able to receive signals from several cells, and the uplink transmission from the UE could be received by several cells. Currently the signals from the other cells would be interference to the UE in LTE, but if there were a way to coordinate the downlink transmissions from all cells within the range, the result would be a greatly improved signal quality. In the uplink fewer changes would ne needed because the signal could already be received by several cells. Therefore, uplink CoMP probably does not need major specification changes, as it is more of an implementation issue (see Figure 11.14).

11.8.1 Downlink CoMP

For the downlink CoMP, three different approaches have been studied by 3GPP:

- Joint processing (JP);
- Coordinated scheduling or beamforming;
- Dynamic point selection (DPS).

Joint processing takes place when the same transmission to a UE is simultaneously transmitted from multiple transmitters. This approach requires tight coordination between transmitters.

Coordinated scheduling or beamforming (CS/CB) is a form of coordination where a UE is transmitting with a single transmission or reception point that is the base station in the serving cell. CS/CB reduces the interference level experienced by a UE terminal by appropriately selecting the beamforming weights of interfering cells to steer the interference away from the UE.

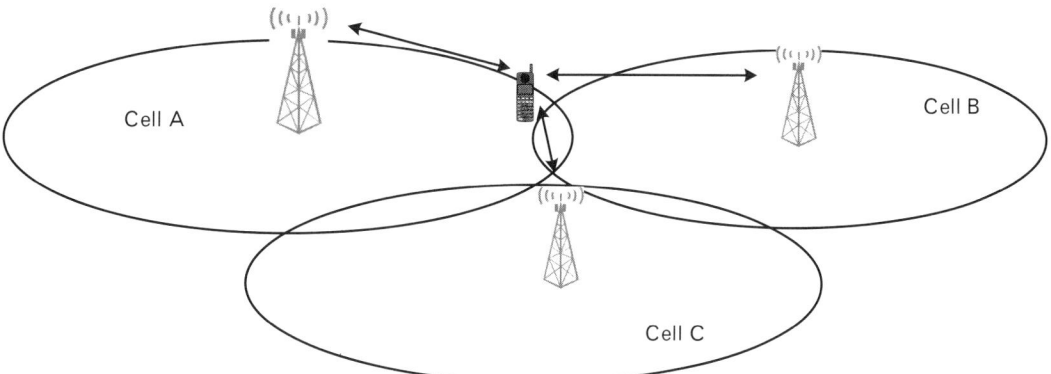

FIGURE 11.14 *Coordinated multipoint transmission/reception.*

In dynamic point selection, the UE is always served by a single transmitter. However, this transmitter may dynamically change, based on the channel state.

All these methods require very fast feedback on channel quality measurements. Also, in case of joint processing and dynamic point selection, the backhaul network will be heavily loaded since the downlink data has to be sent to all potential transmitters. In case of the CS/CB method, only scheduling decisions and details of beams need to be coordinated.

11.8.2 Uplink CoMP

In the uplink direction there are two alternative approaches:

- Joint reception (JR);
- Coordinated scheduling and beamforming (CS/CB).

In joint reception the same signal is received by several receivers at multiple points. The individual signals are then combined and processed to produce the final signal. The main disadvantage with JR is that all participating eNodeBs need to transfer large amounts of data.

Coordinated scheduling operates by coordinating the scheduling decisions among eNodes to minimize interference. This method provides a much reduced load in the backhaul network because only the scheduling data needs to be transferred between the different eNBs.

CoMP mechanisms have been shown to improve signal quality at cell edges. However, because some CoMP mechanisms require extra processing and transmitting data between eNodeBs, they will also increase signal latency, which will be a critical issue for LTE-A. With this in mind, 3GPP has launched a new study in [22] to find out how CoMP would work with nonideal backhaul.

Other CoMP-related 3GPP documents are found from [3, 23].

11.9 LTE and Voice

There is a well-known problem with LTE and its provision of circuit-switched services, the most important of which are the voice calls. Older systems, such as GERAN and UMTS, have both circuit-switched and core-switched domains, and separate entities in the core network can handle either circuit-switched or packet-switched connections. However, LTE has only a packet-switched core network, the evolved packet core (EPC).

When LTE was originally being designed, it was thought that voice could be provided to LTE users via the IP multimedia subsystem (IMS) and thus LTE would not need a circuit-switched domain. This is a good plan, except that mobile operators were not keen on deploying IMS. IMS has been in specifications for about 10 years by now, well before the initial LTE was launched, but it has not been adopted by mobile operators as was originally envisaged. The reasons for this are various, but basically IMS is a major upgrade that requires lots of work: upgrades to equipment, creating the business case for IMS services, interoperability tests, increased signaling and user data because of IMS sessions, and so on. Moreover, there was a lack of IMS-capable devices as phone makers did not bother to add IMS capability into their handsets because there were no IMS-capable networks. And most LTE operators already have 2G/3G networks that can offer circuit-switched services for those who need them. IMS would not bring new services as such, but it could deliver them more efficiently. Therefore, many operators preferred to wait, let the others to do the IMS adoption first, and learn from their experiences/mistakes.

Note that IMS is not (only) about delivering voice over LTE; this is just one IMS service among others. But the lack of voice provision over LTE has forced the operators to do something about the matter. Basically they have three options to choose from:

- Circuit-switched fall back (CSFB);
- Voice over LTE via GAN (VoLGA);
- Voice over LTE.

CSFB is the most common solution operators have adopted. Its principle is that if 4G cannot do voice, then let's do it via 2G or 3G. If the network operator also has a 2G or 3G network, then voice calls are routed via that network using the circuit-switched domain. This feature has been specified in LTE standards since Release 8 (i.e., from the start of LTE), and it has been enhanced since then. However, this functionality is not very "elegant"—it was meant to be a temporary solution for a problem that was expected to go away soon. CSFB includes some challenges, mainly with the call setup times in 2G/3G. When the mobile gets an indication in LTE that there is an incoming CS call (or indeed the mobile user wants to make a call), everything has to happen quickly because the user will not accept if the mobile stays quiet for several seconds before a CS call has been established with the other network.

The first problem to solve is how the mobile gets the indication of an incoming CS call in the first place. If the mobile is camped on the LTE network, the other network (2G or 3G) in principle does not know where the mobile is and thus cannot route the incoming call indication appropriately. LTE employs tracking areas (TAs) to register the location of the mobile, whereas 3G uses location areas (LAs) for the same purpose. The movement

TABLE 11.6 MAPPING TABLE OF LAs AND TAs

LTE Tracking Area	3G Location Area	MSC/VLR
TA1	LA1	MSC/VLR1
TA2	LA4	MSC/VLR4
TA3	LA2	MSC/VLR2
TA4	LA2	MSC/VLR2
TA5	LA3	MSC/VLR3

of the mobile within the LTE network will trigger a tracking area update if the mobile enters a new TA. As a result of this procedure, the new TA identity will be stored in the MME. However, the MSC/VLR in 3G does not have knowledge about this procedure, and, even if it was told about the new TAI, it would have no meaning to MSC/VLR because that 3G register knows only LAIs.

The solution is to combine the mobility management processes of the 3G CS domain and LTE. The MME in the LTE network has to maintain a mapping table of LAs and TAs. This table informs the MME on where to forward (i.e., which MSC/VLR) the location registration message that is received from the mobile. Each TA is coupled with an LA that overlaps its area. Note that this table is operator maintained and has to be renewed every time TAs or LAs are reconfigured.

After the location registration, also the MSC/VLR in 3G receives the LTE location information (i.e., the identity of the MME on which area the mobile is currently located). The MSC/VLR can use this information to relay the circuit-switched call request from 3G to the correct address in LTE.

Once the mobile in the LTE network receives a voice call setup request from 3G, the most common solution is an LTE to UMTS redirection in which the mobile only reads the mandatory system information blocks (SIB) from the new 3G cell before accessing it. An additional challenge is that there are different types of 2G/3G networks; those who need to handle CSFB include GSM, UMTS, and CDMA2000. They all need (more or less) different mechanisms to handle CSFB.

It is not only voice calls that need special treatment in early LTE networks—SMS is another important service that lacks the SMS-over-IMS functionality in those networks. Instead, a temporary solution is available as *SMS-over-SG interface*. This solution enables the LTE network to transmit circuit-switched SMSs. SG is an interface in the core network, between the MME in the EPS (in LTE), and the mobile switching center (MSC) in the CS domain. SG is tasked to handle mobility management and paging procedures, and now, in early LTE systems, also SMS. SMS-over-SG is a different procedure from CSFB. Since SG is an interface within the core network, this procedure is transparent to mobiles. The procedure does not trigger a fallback to 2G/3G, and thus it does not require the presence of those networks as CSFB does. Standards-wise, if the network supports CSFB, it also has to support SMS-over-SG, but this requirement does not apply the other way around.

The story of voice in LTE networks has been a slightly embarrassing episode for the LTE community. Here we have a state-of-the-art telecommunications system that is capable of everything . . . except you cannot make a voice call to your mother! The issue is being addressed, but the most widely used CSFB solution is clearly a temporary one. Its biggest drawback is that an LTE operator also needs overlapping 2G/3G coverage before CSFB can be employed. Whereas this is normally the state of affairs anyway, it is by no means carved into stone. A new entrant operator (so called greenfield operator) does not have a "backup" network into which to make a call fallback. Such an operator may want to implement IMS services in LTE already at the network launch. However, there is still a problem with the supply of IMS capable mobile handsets.

References

[1] 3GPP TR 36.927, v 11.0.0, Evolved Universal Terrestrial Radio Access (E-UTRA); Potential Solutions for Energy Saving for E-UTRAN; 09/2012.

[2] 3GPP TR 36.887, v 0.3.0, Evolved Universal Terrestrial Radio Access (E-UTRA); Study on Energy Saving Enhancement for E-UTRAN; 10/2013.

[3] 3GPP TS 36.300, v 11.7.0, Evolved Universal Terrestrial Radio Access (E-UTRA) and Evolved Universal Terrestrial Radio Access Network (E-UTRAN); Overall Description; Stage 2; 09/2013.

[4] 3GPP TS 36.211, v 11.4.0, Evolved Universal Terrestrial Radio Access (E-UTRA); Physical Channels and Modulation, 09/2013.

[5] 3GPP TS 36.213, v 11.4.0, Evolved Universal Terrestrial Radio Access (E-UTRA); Physical Layer Procedures; 09/2013.

[6] 3GPP TR 36.871, v 11.0.0, Evolved Universal Terrestrial Radio Access (E-UTRA); Downlink Multiple Input Multiple Output (MIMO) Enhancement for LTE-Advanced; 12/2011.

[7] 3GPP TS 36.116, v 11.3.0, Evolved Universal Terrestrial Radio Access (E-UTRA); Relay Radio Transmission and Reception; 07/2013.

[8] 3GPP TS 36.216, v 11.0.0, Evolved Universal Terrestrial Radio Access (E-UTRA); Physical Layer for Relaying Operation; 09/2012.

[9] 3GPP TR 36.806, v 9.0.0, Evolved Universal Terrestrial Radio Access (E-UTRA); Relay Architectures for E-UTRA (LTE-Advanced); 03/2010.

[10] 3GPP TR 36.826, v 11.3.0, Evolved Universal Terrestrial Radio Access (E-UTRA); Relay Radio Transmission and Reception; 07/2013.

[11] 3GPP TS 36.101, v 12.1.0, Evolved Universal Terrestrial Radio Access (E-UTRA); User Equipment (UE) Radio Transmission and Reception; 09/2013.

[12] 3GPP TR 36.808, v 10.1.0, Evolved Universal Terrestrial Radio Access (E-UTRA); Carrier Aggregation; Base Station (BS) Radio Transmission and Reception; 07/2013.

[13] 3GPP TR 36.823, v 11.0.0, Evolved Universal Terrestrial Radio Access (E-UTRA); Carrier Aggregation Enhancements; User Equipment (UE) and Base Station (BS) Radio Transmission and Reception; 03/2013.

[14] 3GPP TS 36.423, v 11.6.0, Evolved Universal Terrestrial Radio Access Network (E-UTRAN); X2 Application Protocol (X2AP); 09/2013.

[15] 3GPP TS 36.440, v 11.2.0, Evolved Universal Terrestrial Radio Access Network (E-UTRAN); General Aspects and Principles for Interfaces Supporting Multimedia Broadcast Multicast Service (MBMS) within E-UTRAN; 03/2013.

[16] 3GPP TS 36.441, v 11.0.0, Evolved Universal Terrestrial Radio Access Network (E-UTRAN); Layer 1 for Interfaces Supporting Multimedia Broadcast Multicast Service (MBMS) within E-UTRAN; 09/2012.

[17] 3GPP TS 36.442, v 11.0.0, Evolved Universal Terrestrial Radio Access Network (E-UTRAN); Signaling Transport for Interfaces Supporting Multimedia Broadcast Multicast Service (MBMS) within E-UTRAN; 09/2012.

[18] 3GPP TS 36.443, v 11.3.0, Evolved Universal Terrestrial Radio Access Network (E-UTRAN); M2 Application Protocol (M2AP); 06/2013.

[19] 3GPP TS 36.444, v 11.6.0, Evolved Universal Terrestrial Radio Access Network (E-UTRAN); M3 Application Protocol (M3AP); 06/2013.

[20] 3GPP TS 36.445, v 11.0.0, Evolved Universal Terrestrial Radio Access Network (E-UTRAN); M1 Data Transport; 09/2012.

[21] 3GPP TR 37.822, v 1.0.1, Study on Next Generation Self-Optimizing Network (SON) for UTRAN and E-UTRAN; 09/2013.

[22] 3GPP TR 36.874, v 1.0.0, Coordinated Multipoint Operation for LTE with Nonideal Backhaul; 09/2013.

[23] 3GPP TR 36.819, v 11.2.0, Coordinated Multipoint Operation for LTE Physical Layer Aspects; 09/2013.

CHAPTER 12

Future Developments

12.1 Introduction

In this chapter we try to have a look into the future. The contents of this chapter are divided into two parts. First we will discuss the near future—what kind of improvements are expected to be introduced to LTE-A specifications within the next two to three years (i.e., in Releases 12 and 13). Note that even if something is added into a LTE release now, it will take at least two to three years before that release is deployed by an operator, and in most cases it will take much longer.

In the second part of this chapter the focus is on the fifth generation system. That is expected to be launched around the year 2020. Various forecasts are predicting that by 2020 the mobile data traffic will increase by a factor of 1000, and a new generation system must be designed to provide performance that matches the expected increase in traffic volume. Also, it is widely suggested that the required performance improvements can be achieved by a combination of three factors:

1. More bandwidth;
2. Massive use of small cells;
3. Further improvements of existing mechanisms.

Even though this list is, in principle, quite correct, it gives a much too simplistic view of the challenges that are awaiting 5G system designers. More about this in Section 12.3.

It must also be stressed here that the gradual improvements introduced in LTE-A, and the 5G improvements discussed in this chapter, are not separate issues. LTE-A has to continuously evolve, release by release, so that it can handle the increased data traffic and other challenges it encounters every year. By 2020, it may have evolved into something that can be labeled as 5G.

12.2 Evolving LTE-A

The way 3GPP develops standards is one of gradual improvement. There are some bigger leaps such as adopting OFDMA as a basis of the LTE air interface, thus replacing WCDMA. But still, every new release brings new features and improves existing ones in order to provide better services to users. This process did not stop when LTE was launched in Release 8, and it did not stop when LTE-A was introduced in Release 10. It still continues, with Release 12 to be frozen in late 2014, and Release 13 possibly 1.5–2 years later.

In this section we will have a look at a few possible future LTE-A enhancements. These features are still at a study phase or at stage 1 specification phase, so it is by no means certain that these features will appear in final stage 3 specifications, or if they do, in what format.

12.2.1 Proximity Services

Proximity services refer to a very old idea. It is a service that has been proposed as an enhancement to most mobile communications systems, but has not found success very often so far. Proximity services are known by many other names, too, such as device-to-device (D2D) communications, and walkie-talkie communications.

The basic idea in proximity services is that two UEs close to each other could communicate directly with each other, with only limited involvement by the mobile network, or in an extreme case even without the mobile network. Such functionality might be useful in some special cases. If the user plane is connected directly from one UE to another, then the network user plane resources are not consumed. However, such functionality would still consume air interface resources. Another scenario is the communication of two UEs, at least one of which is outside the network coverage area. Without proximity services, such a connection would not be possible. On the other hand one could ask how common is such a scenario. Mobile phone networks are already widely deployed, and in any case the communicating UEs should be relatively close to each other (i.e., within a few kilometers, depending on radio conditions) before proximity services could be used. The third, and quite important, use case is public safety. Emergency services typically do not trust public mobile phone networks in their communications because in a major accident the local communications network may also be destroyed. Therefore, these services use specialized systems (such as TETRA) that can support direct connections between handsets. However, if LTE-A were to support direct connections, then LTE-A handsets could also be adopted by emergency services.

Figure 12.1 depicts the basic proximity services scenario whereby the user plane is routed directly between two communicating UEs, but the control plane is routed to the MME as usual.

There are also several other possible architectural scenarios for proximity services, some of which are depicted in Figures 12.2 and 12.3. The communicating UEs can be connected via the local eNodeB (if they are both connected to the same eNodeB). UEs may also be connected to two different PLMNs, with no overlapping coverage, but still close enough to each other so that they could try to have a direct proximity services connection.

One UE can also be outside the network coverage while connected to a UE that is within the coverage area, or both UEs can be outside the service area. The latter scenario is especially interesting for emergency services,

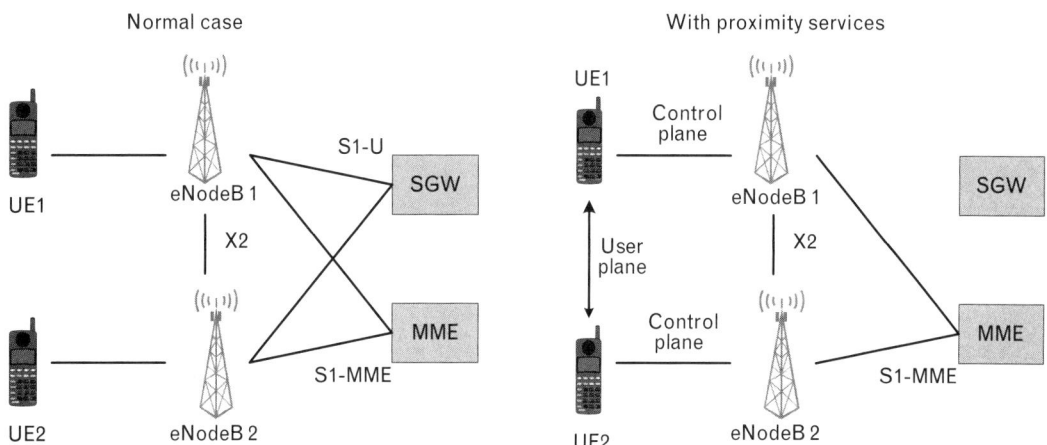

FIGURE 12.1 *Proximity services basic scenario.*

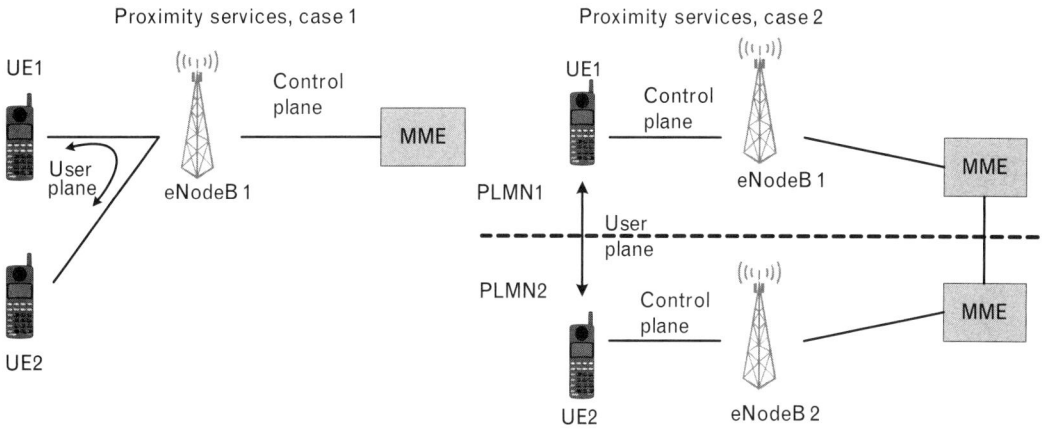

FIGURE 12.2 *Additional proximity services scenarios 1.*

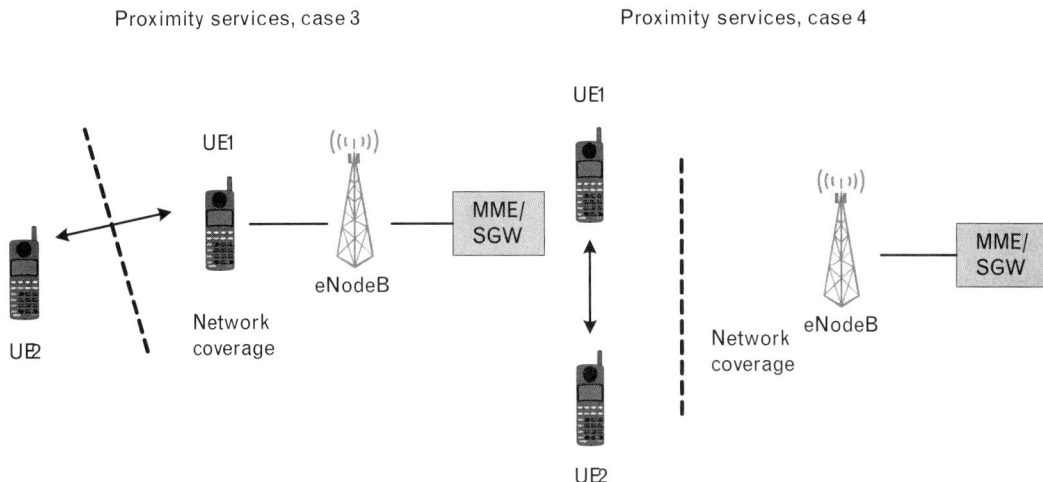

FIGURE 12.3   *Additional proximity services scenarios 2.*

since it would allow proximity service-capable UEs to continue communicating even when the network is not there.

Note that the cases 3 and 4 are problematic in that the UEs' outside network coverage cannot get any control information from the network, at least not directly. Without this radio link configuration information, any connection to another UE is impossible. In case the UEs in question are special UEs that have been designed, say, for emergency services use, they could have been preconfigured so that in case there is no network connection, they would start using a stored configuration that allows them to connect to other UEs in the same preconfigured group. Alternatively, one proximity service UE in a group could be configured to act as a radio resource controller—in a way, a mobile mini-MME (see Figure 12.4). This UE could manage the radio connection control for all UEs within the group.

Another variation of the case in Figure 12.4 is presented in Figure 12.5. In this example the radio resource controller UE also acts as a relay (i.e., the user plane is also relayed via it). In this case the UEs do not connect directly to each other. In case the relay UE itself is within the mobile network coverage, it can also relay UE connections to the PLMN if that is required. In the second case, the relay UE can also relay control signaling to the eNodeB, so it does not need to act as a radio resource controller anymore. In fact, if a radio resource controller-capable UE detects a mobile network, it probably should not allocate radio resources by itself anymore but relay such requests to this network. This is because any radio resource allocation done by it may clash with radio resource allocations done by the mobile network control node (i.e., the MME).

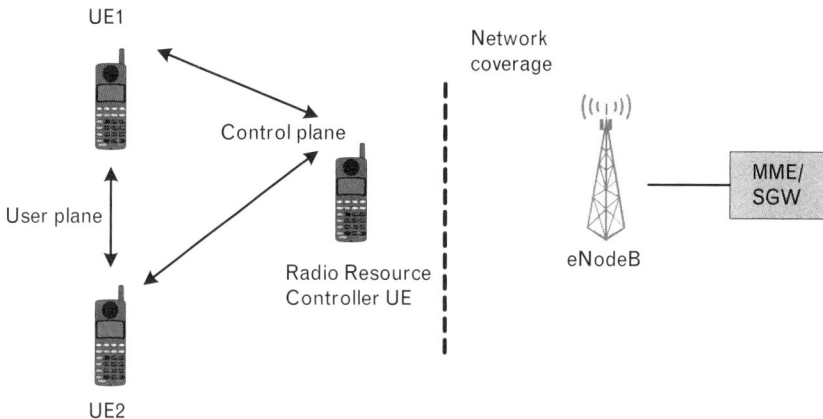

FIGURE 12.4
Emergency services radio resource controller.

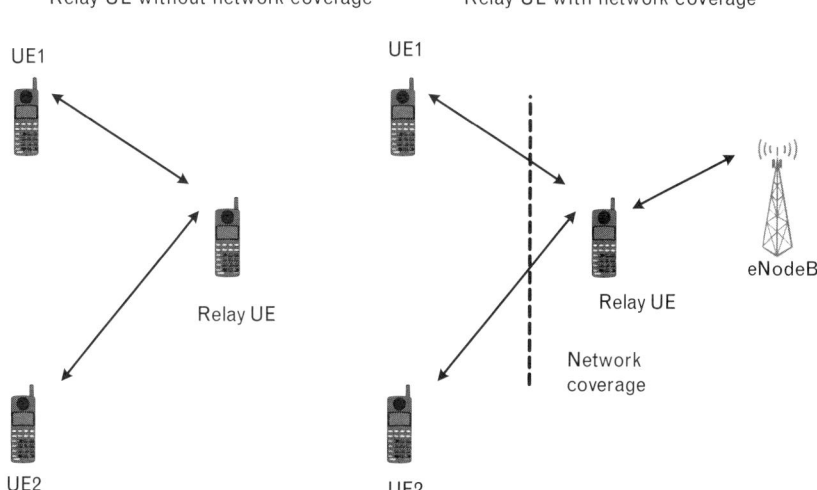

FIGURE 12.5
Relay UE with and without network coverage.

Proximity services would also allow a kind of closed user group communication whereby the members of a group (members of a family, club, company, and so on) could communicate directly with each other.

Proximity services are a fundamentally new concept for LTE-A. Therefore, introducing such services is not an easy task. Many new mechanisms would need to be introduced in specifications. Most of them are relatively easy to solve technically; rather, the problems are more at the policy level. For example, to be able to form a direct connection with another UE, the user must first know if the other UE is nearby. The process of finding out who is nearby is called proximity discovery. Several proposals exist on how this could be solved [1]. However, proximity discovery would clearly breach other user's privacy by revealing their location information. Issues like this have to be solved by new proximity services policies or rules.

Finally, even though all technical issues around proximity services are possible to solve, the main question is whether there is a strong enough business case behind it. New features to mobile networks are introduced by mobile operators. They are investing real money to launch new services and expect to get their investments back within limited number of years. In case of proximity services, it is not clear what the benefit would be for operators (i.e., how can they make money out of this). They could of course charge for the privilege of being a member of a closed user group that can use the proximity services. But then again, why would anybody want to pay extra for something that they can already do without proximity services? Use cases where one or more UEs are outside the network coverage are extremely rare. For the author of this book, the only solid use case is the use of proximity services by emergency services. However, it is not clear if this use case is enough for network operators to make an investment decision regarding proximity services.

Proximity services are still in a study phase in 3GPP. Some study results are already available in [1, 2].

12.2.2 Machine-Type Communications

Machine-type communication (MTC) is data communication that involves entities that are not human. Its characteristics are very different from the usual human-to-human communication, at least in following aspects:

- Different market scenarios;
- All communication is data;
- Communication has to be very low cost;
- Very large number of communicating devices;
- To a large extent, little traffic per device.

MTC devices and services do not form a single coherent group. Certain MTC services are more tolerant and can accept a lower level of performance (e.g., longer latencies and low data throughput). However, other MTC services will have similar service requirements as current LTE-A mobile network services.

The 3GPP technical report [3] has a list of examples of MTC applications (see Table 12.1).

The work to add support for MTC is still at early study phases in 3GPP. Much more will be needed before suitable solutions are found. However, MTC support is an important LTE-A enhancement that has to find its way to 3GPP specifications in the coming years. The problem is that MTC applications are so varying that each application class may require its own solutions.

TABLE 12.1 Possible MTC Applications

Service Area	MTC Applications
Security	Surveillance systems Backup for landline Control of physical access (e.g., to buildings) Car/driver security
Tracking and Tracing	Fleet management Order management Pay as you drive Asset tracking Navigation Traffic information Road tolling Road traffic optimization/steering
Payment	Point of sales Vending machines Gaming machines
Health	Monitoring vital signs Supporting the aged or handicapped Web access telemedicine points Remote diagnostics
Remote maintenance/control	Sensors Lighting Pumps Valves Elevator control Vending machine control Vehicle diagnostics
Metering	Power Gas Water Heating Grid control Industrial metering
Consumer devices	Digital photo frame Digital camera eBook

The following sections contain some examples of MTC applications and problems that need to be solved in each case [3].

Addressing from a Centralized Entity
Metering devices can be controlled by a centralized entity that informs the actual metering device when it needs measurement results. In some cases, there may be a requirement for a low latency response.

Theft-Vulnerable MTC Device
MTC devices are often installed in remote locations that are not monitored by humans. Therefore, they are prone to vandalism or theft. The network

cannot prevent this from happening, but it should be able to detect what has been done to the device in order to deactivate the device and the related USIM.

Time-Controlled MTC Device
Some MTC applications are not time critical. The communication can take place at any time. A good example of this is an electric power meter reading. The network operator can configure these types of devices to communicate only during low traffic periods.

Avoid Radio Network Congestion
Some MTC applications tend to generate bursty traffic. These are typically monitoring sensors. When a triggering event occurs, it tends to occur for a large number of MTC devices simultaneously. The radio network must be prepared for a large number of MTC devices in a particular area to transmit data almost simultaneously.

Avoid Core Network Congestion
The radio access network may be able to handle the bursty traffic case described in the previous paragraph (e.g., when MTC devices are well scattered over the network). Despite this, when a large number of MTC devices is transmitting simultaneously, congestion may take place in the core network or on the link between the core network and the MTC server.

Avoid Signaling Network Congestion
The previous two examples were about congestion caused by bursty payload data traffic. However, MTC devices may also generate bursty signaling traffic, which may cause problems. For example, after a power outage, some MTC devices may reboot themselves at the same time and then send almost simultaneous messages to the MTC server to inform the server about their active status.

Access Control with Billing Plan
An USIM should only be used with the associated MTC device, and there should be a mechanism to enforce this. Many MTC devices have very low cost billing plans, and therefore using such an USIM with other non-MTC devices is not acceptable.

Extra Low Power Consumption Devices
There are many MTC use cases where extra low power consumption is essential because the power source for the MTC device cannot be replaced (or the replacement is so difficult that it should only be done in extreme cases). These devices include, for example, many battery-powered tracking devices

and battery-powered sensors. That is, the system must be designed so that extra low power consumption in communication is possible.

Extra Low Power Consumption with Time-Controlled MTC Devices
An MTC device may be configured to communicate within predefined period and additionally be prepared to receive nonperiodic messages outside these periods. This case is a variation of the previous use case. Outside the defined communication periods, it should be possible for the network operator to configure the MTC device to enter a sleep mode, but there should also be a mechanism in place to wake the device in case an urgent communication need arises.

End-to-End Security for Roaming MTC Devices
For some MTC applications security may be important. If an MTC device is roaming outside the home network coverage area (an MTC device can be mobile or it might have been installed outside the home network in the first place), then the MTC application server cannot trust that the communication is secure. In this kind of case, it may be necessary to provide end-to-end security for messages exchanged between the MTC server and devices that are roaming.

Machine type communications is discussed, for example, in [3]. This is a stage 1 specification, so there is still a long way to go before stage 3 specifications are available.

12.2.3 Mobile Relays

An increasing number of countries either have operational high-speed train systems or have plans to build such systems. In East Asia, Japan, China, Korea, and Taiwan all have operational bullet train networks, and high-speed train networks are also spreading across Europe. Such trains provide special challenges to mobile networks. In this section, we will discuss these challenges and possible solutions.

High-speed train challenges for mobile communications include the following:

1. High-speed vehicles were not taken into consideration when traditional cellular networks were first launched.
2. Train carriages typically suffer from poor signal penetration.
3. High-speed rail tracks have lots of tunnels in some mountainous countries.

High-speed trains are fast, but the speed itself is not a major problem for LTE devices since they can communicate with the network even when moving very fast (one of the design requirements of LTE-A was the ability to have connections at 500 kph). The fastest high-speed trains can move at speeds of about 320 kph in normal operations. This speed translates to about 90 meters per second. The base station coverage area can be highly varying, but typically those are only a few hundred meters in diameter. This would mean that a train enters a new cell area every few seconds. Moreover, the cell overlap area where the handover has to be performed is even shorter. Since the train is fully packed with passengers, there would be very many simultaneous handovers very frequently. Even if a mobile device is not in an active mode, it may still communicate its new location to the network once it changes a cell (i.e., tracking area update). The signaling load generated to the network can potentially be very large, and frequent cell changes will drain UE batteries much faster than from stationary devices (see Figure 12.6).

Doppler frequency shift may also become a problem on high-speed trains. It will cause problems with the signal quality.

Train carriages are made of metal, which is not good for mobile phone signals. Even the carriage windows are typically reinforced with metal mesh in high-speed trains. That is, the signal penetration inside the train may be poor.

Tunnels present particular problems to wireless communication, and some mountainous countries such as Japan and Korea have lots of tunnels along their high-speed rail tracks. Tunnel entrances in particular are problematic since there the cell overlap area is small which means that there is very little time to perform a large number of handovers. Inside the tunnels normal macro eNBs are mostly useless if the train does not have outside antennas. This is because signal penetration losses with low incidence angles are very high, and in tunnels these angles would most of the time be close to zero. Radiating cables or remote radio units (RRUs) can be used inside tunnels to provide coverage, but the entrances of tunnels are still problematic.

While it is possible to alleviate these issues with careful network planning to continue using existing technologies in high-speed trains, there are now proposals to adopt mobile relays to solve the problems. Mobile relay is

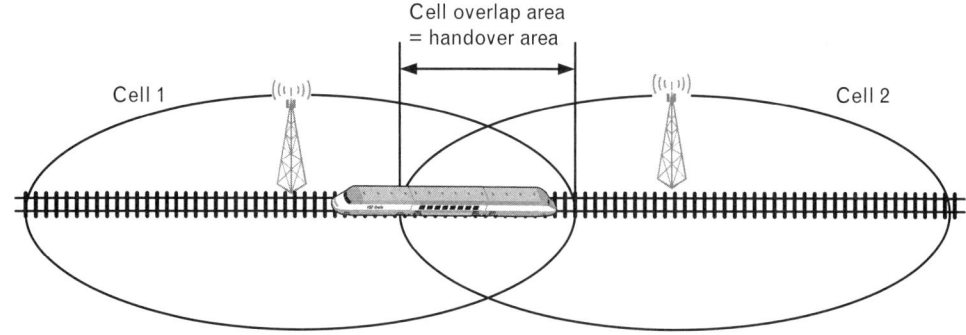

FIGURE 12.6 *High-speed train and the handover problem.*

a relay mounted on a vehicle and wirelessly connected to macro cells. It is likely to provide at least the following key functions:

1. Wireless connectivity to end users inside the vehicle;
2. Wireless backhaul connection to the network outside the train;
3. Capability to perform group mobility functions;
4. Capability to allow different air interface technologies on the backhaul and on the access link.

It must be noted that mobile relays are not yet specified for LTE-A. A feasibility study on mobile relays has been ongoing during Release 11, and the work may continue in Release 12. Several alternative solutions have been presented, the difference of which is mainly on which parts of the fixed relay architecture are modified in order to provide support for mobility.

The advantages of mobile relays are as follows:

- Since the relay nodes (access points) are within the train/other vehicle, there are no handovers/tracking area updates when the train moves;
- Signals do not suffer from high penetration losses;
- No Doppler effect;
- Reduced UE power consumption.

Instead of UE handovers and tracking area updates, the mobile relay itself will perform a handover/tracking area update (but these are transparent to UEs in the train). This procedure is also known as a group handover.

Since the UEs and relays nodes are in same carriages, there is no signal penetration loss, nor there is Doppler effect. All these issues will result in reduced UE power consumption.

And as always there are also disadvantages:

- Increased cost;
- Increased latency;
- Standardization effort;
- Unclear ownership of the mobile relay infrastructure.

Mobile relays will have a cost effect. The relay equipment itself may be cheap, but its deployment will cost serious money. Planning the deployment, modifying core network elements (depending on the chosen architecture), taking trains out of service for equipment installation, testing, and so on—it all adds up.

Mobile relays will also increase latency when compared to nonrelay network scenarios. This is true for all relays, mobile or fixed. On the other hand, signal quality will be better and therefore in some scenarios the latency experienced may in fact be smaller because protocols do not need to perform as many retransmissions when relays are used.

The standardization effort also has an effect. It will have a cost effect, and since all 3GPP standardization work is congested (for each release, there are many more work item proposals than can be accommodated), every new feature to be standardized will mean that some other feature will not be standardized.

It is also unclear who owns the mobile relay. Is it the mobile network operator or the train operator? If it is the train operator, does the train operator also need a mobile phone operator license? Should the same mobile relay infrastructure support several mobile operators (network sharing)? These questions (and many others) are all issues that are not really technical but rather commercial or regulatory issues, but they still have to be solved or there will be no business case for mobile relays.

Mobile relays may support multi-RAT functionalities (i.e., an LTE-A Un may provide the backhaul link, while different air interface technologies, such as LTE/3G/2G/WiFi, may be supported on the access link).

As mentioned earlier, the problems described above can be solved or at least alleviated to a certain extend by existing methods such as directional macro cells or RRUs along the train line. Figure 12.7 shows a solution whereby directional macro cells are combined with carrier aggregation (though of course the carrier aggregation part can only be exploited by UEs that support carrier aggregation. Mobile relays would have the advantage that they are transparent to UEs (i.e., any old UE could attach to them). The carriers in freq 1 and freq 2 are configured to UE as component carriers. At any point, the UE can receive good SINR at least from one carrier. This solution increases the overlap area to the cell radius, and if cells are

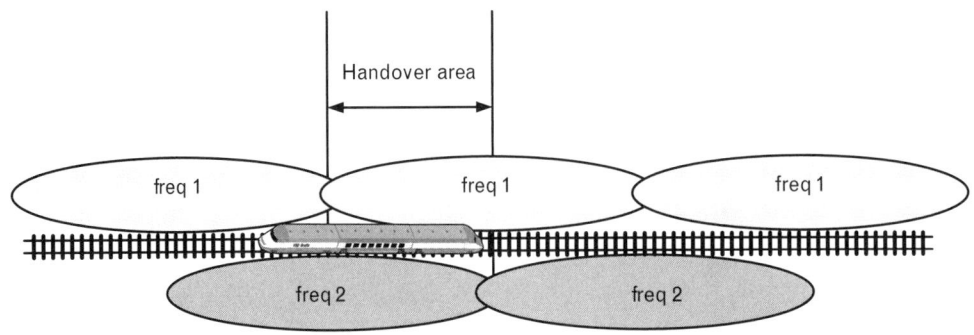

FIGURE 12.7 *Directional macro cells with carrier aggregation.*

directed then the cell radius can be very large. This method could decrease the handover failure rate considerably.

Another solution using existing techniques is to use RRUs as depicted in Figure 12.8. Different remote radio units (RRUs) share the same cell ID. The nearest RRU is chosen to serve the train without the need for handovers. This scheme will enlarge the cell radius together with the overlap area. In fact, the whole train line could be defined as one logical cell. This method is already used in many underground systems; in a tunnel it is usually implemented using a radiating cable.

Mobile relays technical report can be found in [4].

12.2.4 Heterogeneous Networks

Heterogeneous networks (HetNets) are not a new invention as such, but HetNets are becoming very important in LTE-A and that is why they are discussed here. HetNets and related issues are mentioned in several places in this book, but here everything is tied together.

HetNets refer to a network deployment that consists of a mixture of macro, pico, and femto cells (i.e., cells of different sizes). HetNets typically provide layered coverage in which macro cells provide the wide-area coverage and smaller cells provide the capacity in traffic hotspots (see Figure 12.9). Therefore, these cell types are also known as coverage cells and capacity cells.

Mobile data traffic is forecast to grow very rapidly. The current situation in which macro cells are the norm and small cells are added to a few locations will have to change fundamentally. The network capacity has to grow so rapidly that pico and femto cells have to be deployed on a large scale, and they will become the most common cell type.

Both macro and small cells are needed in an operational network. They fulfill different tasks and thus complement each other. Combining macro

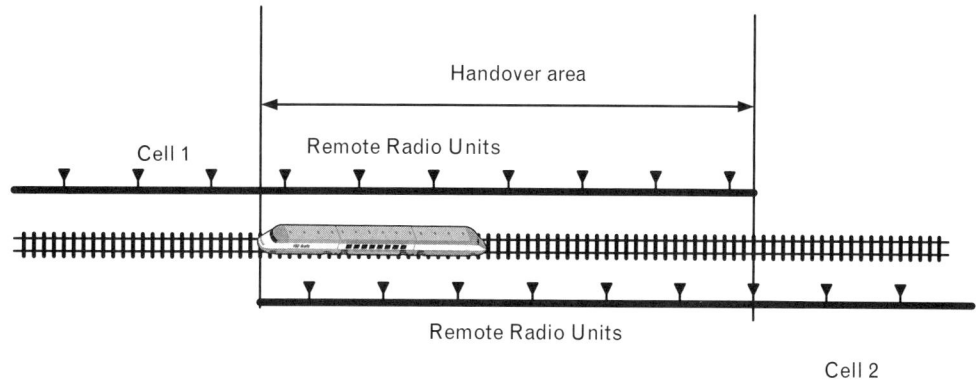

FIGURE 12.8 *Directional macro cells with carrier aggregation.*

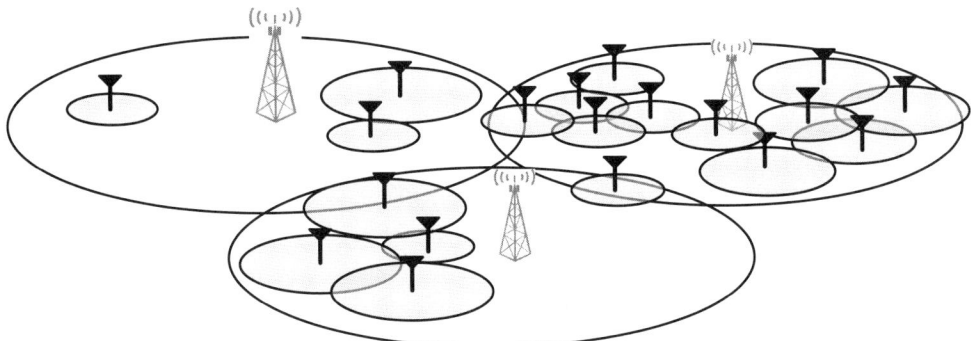

FIGURE 12.9 *Heterogeneous network.*

and small cell deployments can be done in many different ways. The easiest way is to deploy both layers on their own frequency channels, thus avoiding intercell interference. However, spectrum is very expensive and extra spectrum may not even be available. Even if an operator has two frequency bands in its use, assigning a whole layer to a macro cell can be considered (at least partly) waste of resources. Numerous small cells can provide more capacity per area than a few macro cells.

Carrier aggregation is a promising solution to HetNet deployment problems. As discussed in Section 11.4, a macro cell can form one component carrier and a set of small cells another. The advantage over the old setup where we had two "independent" frequency carriers for macro and small cells is that a carrier aggregate-capable UE can swap between component carriers at will, so there is no need to make frequency handovers.

One component carrier will act as a primary component carrier (PCC); others are secondary component carriers (SCC). The control information for all component carriers is sent via the PCC. This arrangement makes carrier aggregation even more attractive to HetNets since it enables the UE to read its control information from a macro cell and receive the user data from a small cell. Note that NTT DoCoMo has promoted similar idea under the label "Phantom Cell" [5].

Carrier aggregation in HetNets would also help to alleviate intercell interference problems that can be severe. If both the macro and small cells operate on the same frequency, then the macro cell may cause serious interference problems to the small cells because typically a macro cell covers small cells completely. When carrier aggregation is used, then the component carriers are on different frequencies, and there is no intercell interference—at least not between different component carriers. However, there are several noncarrier aggregation interference cancellation mechanisms standarized as a result of ICIC, eICIC, and FeICIC work items. These are needed because not all UEs will be carrier aggregation-capable. Older Release 8 and Release 9 UEs will be around for a long time. Moreover, not all HetNets support carrier aggregation. For example, HeNBs are deployed semi-independently from operators, and they will not be part of any carrier aggregation scheme.

Network energy saving is also important in HetNets. Because small cells are typically capacity cells, they will not be needed all the time. On low-traffic periods, some small cells could be switched off, at least if there is a coverage cell nearby making sure that there will not be coverage holes after switch-off. Because small cells are expected to be deployed in large numbers, efficient energy saving schemes would result in considerable savings. Such schemes are not yet standardized, but Release 12 may bring a change to that.

12.3 The Fifth Generation

12.3.1 Introduction

As discussed in the beginning of this book, a new generation in mobile communications technology has been launched about every 10 years so far. And there is no reason to believe that this trend would not continue in the future. This conclusion will thus let us expect the 5G to emerge around 2020. Indeed, we have already heard news from Japan that there will be a 5G system in use when the Olympic games are held in Tokyo in 2020. Japan is slightly ahead of other countries when it comes to adopting new mobile communication technologies so in that sense a promise of 5G is not very far-fetched.

However, there are no 5G standards available as of now—not even serious plans. Various telecommunication businesses are of course studying future technologies as they always do, and in their long-term research they are certainly trying to forecast what is going to happen in 2020. But no international standardization body is preparing 5G specifications yet. And, most importantly, ITU has not defined what 5G means or what it should contain.

It is good to be cautions with the term 5G. It is becoming the new buzzword in mobile telecommunications, and if you want publicity and attention to your new invention, the easiest way to get attention is to call it part of 5G. Therefore it is good to take the news about 5G with a pinch of salt, especially if they come from a less-known source.

In general, a new mobile telecommunications system must be able to provide more, better, and faster everything. For example, users want higher data rates and shorter latencies, and operators want more capacity, cheaper equipment, and easier-to-operate networks. The wish list for 5G could include:

- More capacity;

- Higher user data rates (especially average data rates);
- Shorter latencies.

Note that bullet points 1 and 2 do not mean the same. More capacity in the network may result in higher user data rates, but not necessarily. Adding more femtocells to a network will increase its capacity, but not necessarily the data rates a user is experiencing. There are lots of different methods to improve the performance of mobile networks, eventually leading to 5G. In this section, we will discuss a selection of them. In any case, the new 5G is not achieved only by improving one single issue. For example, upgrading the air interface access technology in 2020 will bring improvements in performance, but this change will not be big enough so that the new system could provide performance and service expected from a 5G system. A 5G system will be a result of many large and small improvements, and together those improvements will hopefully provide a system that fulfills the expectations for a 5G system.

The list of improvements that are needed for a 5G include at least the following ones:

- New air interface;
- More bandwidth;
- Updated network architecture;
- Widespread use of small cells;
- Use of multiple antennas;
- Support for M2M communication.

These issues are discussed in following sections.

12.3.2 New Air Interface

Generally speaking, the capacity bottleneck of a mobile telecommunications system is typically considered to be the air interface. The main goal behind the development of new air interface technologies has always been to provide more capacity. However, it is wrong to say that capacity can be increased only by adopting a new air interface technology. There are several methods available for the network operator to increase the capacity. Which ones are chosen depend on the particular operator and the particular circumstances.

A new air interface technology can bring improvements to system capacity. The technology changes from TDMA to WCDMA to OFDMA have been leaps in terms of performance. The reason new technologies have been adopted about every 10 years is mainly due to improvements in microprocessor technology. As Moore's Law states, the number of transistors on integrated circuits doubles approximately every two years. One consequence of this is that the chip performance doubles every 18 months. This may not sound a lot, but in 10 years the chip performance increases about hundredfold. Because of the much increased processing capability, the receivers can employ more complex receiver algorithms that were impractical in the past, and consequently new technologies can be adopted.

Recently there has been increased interest toward nonorthogonal multiple access (NOMA) schemes. LTE uses OFDMA, which is an orthogonal access scheme. In orthogonal access schemes the signals from/to different users are orthogonal to each other. The signals do not interfere because their cross-correlation is zero. Orthogonality can be achieved by TDMA, FDMA, or OFDMA access schemes. However, CDMA is a nonorthogonal access scheme, including the WCDMA scheme that is employed in UMTS. NOMA schemes rely on interference cancellation algorithms to sort out the correct signals. Such algorithms have already been used in base stations, but in 2020 it will be possible to employ them also in UEs.

The principal difference between orthogonal and nonorthogonal multiple access schemes is that in orthogonal schemes the channel resource is allocated exclusively to one user only at a time, whereas in nonorthogonal schemes this resource belongs to all active users. Therefore, nonorthogonal access schemes can have a higher overall capacity because the whole capacity of the cell can be accessed by all users. In orthogonal schemes the capacity is divided into small portions that are each allocated to one user only. If a UE cannot use the whole capacity that is allocated to it, the leftover capacity remains unused.

A NOMA scheme can also be used together with an orthogonal scheme. For example, a block of OFDMA subcarriers could be jointly allocated to a group of users, but within that subcarrier block the channel resources are accessed using a NOMA scheme.

12.3.3 Bandwidth

More bandwidth equals more capacity. However, bandwidth is a very limited resource and in most markets it is therefore expensive to acquire. Radio regulators favor spectrum auctions in which the market can decide the price for the bandwidth. These prices tend to be high, and spectrum price may become an insurmountable obstacle for new operator candidates.

LTE-advanced can employ many different spectrum bandwidths—it is a truly scalable system. Currently (in Release 11) LTE-A air interface supports six different channel bandwidths, ranging from 1.4 MHz to 20 MHz. Mobile devices can then use these channels in various combinations, depending on their capabilities and bandwidth requirements. With carrier aggregation, it is possible, in theory, to use 100 MHz of LTE spectrum for one UE.

However, it is expected that the demand for mobile data capacity will increase a thousandfold by 2020. This kind of increase can only be accommodated with the help of several different improvements, one of them definitely being more bandwidth.

It is difficult to find large junks of unused spectrum from the traditional bands used by mobile networks. So called digital dividend is the spectrum that was/is freed up after analog television broadcasting switched or switches to digital transmission. Part of this spectrum will be available for mobile networks, but the problem is that analog TV spectrum was quite low-frequency (usually VHF = 174 to 230 MHz and UHF = 470 to 862 MHz), and therefore only a part of UHF can be considered for mobile broadband. And even though this spectrum is, in principle, very desirable for mobile use because low-frequency signals can penetrate obstacles such as walls and can be used in cells with large coverage, it is not very useful for 5G. In 5G one of the design principles is to deploy a large number of small cells to increase system capacity. Low frequencies are not good for this purpose because they easily carry signals too far, causing intercell interference. These frequencies can be useful for 3G and 4G networks, though.

Instead 5G systems could use spectrum from very high frequencies such as 10 GHz or even 20 GHz. The signal penetration is then poor and cell sizes are small, but this is actually desirable in 5G. Only with very small cells can the overall system capacity be increased enough.

High frequencies have also another desirable property: they make the design of large MIMO antenna installations easier. Efficient MIMO antenna elements at the UE typically have to have an antenna separation of at least 0.5λ before acceptable spatial diversity can be achieved. Wavelength of course changes with the frequency as indicated by (12.1):

$$\lambda = c/f \qquad (12.1)$$

where c = the speed of light and f = frequency.

The minimum separation of MIMO antennas for selected frequencies is given in Table 12.2.

The results from Table 12.2 indicate that for under 1-GHz spectrum bands it is difficult to design a handset with MIMO capability. However, for over 2-GHz spectrum bands this is rather easy, and for 10- and 20-GHz bands it is possible to add a large number of MIMO antennas into even a small handheld device. At this point it is good to note that the half-wavelength rule can possibly be circumvented in the future with new advanced antenna designs. Moreover, the high data rates requiring MIMO antennas are typically associated with large form factor devices such as laptops and tablets, and those can more easily accommodate MIMO antennas even in

12.3 THE FIFTH GENERATION   261

TABLE 12.2 WAVELENGTHS AND FREQUENCIES

WAVELENGTH/2	FREQUENCY
22 cm	700 MHz
17 cm	900 MHz
8.3 cm	1800 MHz
6.2 cm	2.4 GHz
4.28 cm	3.5 GHz
1.50 cm	10 GHz
0.75 cm	20 GHz

relatively low frequencies. A small handheld smart phone, even if used for watching streaming video, does not consume as much bandwidth as a large laptop with a high-resolution display.

The World Radio Conference 15 (WRC-15) will be a very important occasion regarding future radio spectrum allocations, and therefore it will also have a big impact on 5G. WRCs are organized by ITU and historically they have been held every two to five years. The next WRC will be held in Geneva, 2-27.11.2015. Note that "15" in the conference name does not mean that this is the 15th WRC conference. Rather it refers to the conference year. The previous WRC conference was WRC-12. Among other things, the WRC will decide which new radio frequencies can be used for future mobile broadband systems. Since so much is at stakes in this meeting, the preparation phase for it is very long. For example, for WRC-15 the first regional preparatory meetings were held already in 2012, more than three years before the actual conference.

12.3.4 Network Architecture

The new 5G system may also introduce something new for mobile network architecture. In 4G, the architecture was already greatly updated from 3G, with the new access network EUTRAN and the new core network EPC. But 5G faces new challenges. With the expected increase of data traffic and the demand for ever-shorter latencies, a new architectural upgrade will be necessary.

The most obvious solution to accommodate the expected increase in data traffic is to reduce cell sizes; add many more pico and femto cells, and increase the use of SIPTO-like mechanisms that can offload the traffic to fixed data networks as early as possible. More small cells covering a given area can handle more data than a few large macrocells. However, macro cells are also needed to provide large area coverage and also to handle fast-moving UEs, which could cause severe problems in small cells with their frequent handovers. To make small cell deployments more effective, NTT

DoCoMo has proposed a mechanism called phantom cells [5] (see Figure 12.10), whereby the C-plane for small cells is provided by the macro cell covering the area, and the small cells only transfer the U-plane data. The macro cell and small cells would operate on different frequency channels, and the macro cell would also take care of fast-moving UEs. This kind of arrangement also makes network energy saving easier: small cells are likely to have only a few UEs in them. If the network notices that the number of users gets very low, or there are no users at all in a given cell, it is easy to switch that cell off. The control signaling will be handled via the macro cell anyway. This is useful because UEs in a dormant small cell area are not in a coverage hole in this scenario. They can send their connection requests via the macro cell as normal and can be served via the macro cell user plane. If there are enough active UEs in an area, the network can switch the dormant small cell on. Therefore, it can be concluded that smaller cell sizes increase the granularity of the network topology and thus make it easier to switch off lightly loaded cells.

Shorter latencies are another "standard" requirement for every new mobile communications generation. It is also a difficult or at least an expensive requirement to fulfill because it typically requires changes to network architecture. In LTE this requirement was handled, for example, by the flat RAN architecture. In 5G the already short latencies of 4G can be shortened by using fast backhaul technologies (e.g., optical cables) and using user plane data offloading whenever possible. That is, if it is not prevented for other reasons, the user plane data should be offloaded already from the local switch to the fixed data network.

12.3.5 Multiple Antennas

Multiple antenna techniques are known to improve the data throughput considerably. They are especially attractive in the sense that they do not require extra resources such as more bandwidth or more cell sites. Multiple antennas can be exploited in many ways, but possibly the most promising

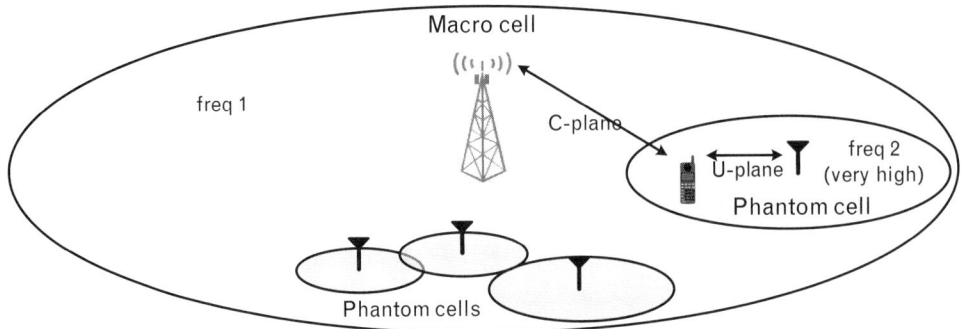

FIGURE 12.10   *Phantom cells.*

techniques with regard to 5G are MU-MIMO and beamforming. MU-MIMO was already discussed in Chapter 11.

Beamforming works better the more antennas there are at the transmitter. In LTE (Release 11) the number of antennas is limited to eight. Increasing this number is possible in future LTE specifications, but the problem is that in currently used LTE frequency bands, the physical size of MIMO antenna installations is getting large because of the required minimum antenna separation. With the new expected high-frequency bands (3.5 GHz and up), it is possible to position the individual antenna elements very close to each other. This enables the use of several tens (or even hundreds) of antenna elements together. This kind of system is known as massive MIMO in many academic papers [6]. Massive MIMO antennas at the base station enable the use beamforming with very narrow beams. Because the beam energy will be concentrated on a small area, this reduces the interference caused to others and improves the signal quality near cell edges.

Indeed, beamforming as such is an old technique, but its usefulness was limited by the number of individual beams a base station could emit. With massive MIMO, it is possible to create individual beams for each user.

Still, it is good to recognize that massive MIMO is not yet ready in the sense that there are many problems to be solved before this scheme can be used in operational networks.

12.3.6 Support for M2M Communications

The number of human mobile network users is not growing anymore in mature markets (though their use of data services is growing rapidly). However, the number of devices that wish to communicate wirelessly is growing fast. This machine-to-machine (M2M) communication (also known as MTC) is very varying in nature; it can include various sensors, alarm systems, meter readings, vehicle-to-vehicle communications, and so on. Some of these devices already use mobile networks for their communications, but especially with low-power sensors the solutions so far have been proprietary. It would greatly improve the usability of these devices if they could communicate using standardized interfaces and protocols. They would also be cheaper and easier to deploy if their connectivity was provided by a standardized solution. Currently, a simple sensor may cost only a few dollars, but its deployment and testing may cost hundreds of dollars. A 5G-connnected sensor would ideally be a plug-and-play product.

The problem with connecting a machine such as a sensor into a mobile network is that the requirements for such a connection are often very different from the UE requirements. For example, a 5G UE will require very high data rates and low latency. It also has considerable processing power in its use, and its battery will be recharged regularly.

On the other hand, a wireless sensor typically transmits only a very small amount of data and it does this very rarely. A sensor has only limited data-processing capabilities and even more limited power sources. Using the same mobile network to cater to the needs for both traditional UEs and tiny sensors will be difficult. Furthermore, M2M communication characteristics are very different by nature; it is not possible to define M2M communication as a single class and then define a solution for this class. For example, a humidity sensor in the garden and a networked car have completely different communications needs.

In the future there will be a very large number of sensors or other M2M communication devices deployed. Even though the amount of data each device will transmit may be small and infrequent, it can be foreseen that those traffic bursts can be simultaneous and this may cause problems with network access. There will be lots of sensors deployed, many of them performing the same tasks. Therefore, once a triggering event happens, it may trigger the transmission of a signal from a large amount of sensors simultaneously.

Due to the nature of M2M devices, the network access protocol has to be lightweight and require as little processing as possible at the device.

The challenges with M2M communication are difficult, but solving them would bring great benefits. And even if purely technical issues are solved, there is still the question of a suitable business case. How would a mobile operator benefit if its network is open for M2M devices? How would it charge the sensors for the use of the network? How would it check that the sensor accessing the network is legitimate? A SIM card and a full authentication protocol may be too heavy a solution for a cheap sensor, but is there an alternative?

12.3.7 Other Improvements

Indeed, 5G should exploit all enhancements available to increase the performance of the system. Every little thing helps to increase the system capability, and the accumulated effect of even small individual improvements can be substantial. For example, the introduction of 256QAM modulation might help in some special cases, such as in indoor femtocells.

Finally, it must be stressed that having something made possible technology-wise is not enough if the new feature is not attractive business-wise. Most new features and enhancements mentioned here are clearly beneficial for a network operator, and therefore it is in operator's interest to adopt them. However, there are also cases where the technical benefit is obvious, but it is not so obvious whether the entity making the investment will benefit. For example, a HeNB is technically a great invention and benefits the operator. But it should be acquired by a homeowner. Therefore, an operator should plan its HeNB business case in a way that benefits all parties involved. Similar cases will no doubt emerge also with M2M communications. The history of mobile telecommunications is full of great inventions that were never deployed. To be able to succeed, a technical solution needs also a credible business case to support it.

12.4　METIS Project

By now, the 5G wagon has definitely started to roll. In Europe, the EU is funding a large project called mobile and wireless communications enablers for the twenty-twenty information society (METIS). Its task is to study the future requirements for mobile communications and set the foundations for 5G. This project was launched in November 2012 [7], and it has already outlined the first set of requirements for 5G [8]:

- Ten to 100 times higher typical user data rates, where in a dense urban environment the typical user data rates will range from 1 to 10 Gbps;
- One thousand times more mobile data per area (per user) where the volume per area (per user) will be over 100 Gbps/km2 (resp. 500 GB/user/month);
- Ten to 100 times more connected devices;
- Ten times longer battery life for low-power massive machine communications where machines such as sensors or pagers will have a battery life of a decade;
- Support of ultrafast application response times (e.g., for tactile Internet) where the end-to-end latency will be less than 5 ms with high reliability;
- A key challenge will be to fulfill the previous requirements under a similar cost and energy dissipation per area as in today's cellular systems.

METIS must be taken seriously because it has an impressive list of members, which includes equipment manufacturers, telecommunications operators, and academia.

Equipment Vendors

- Alcatel-Lucent
- Anite
- Ericsson
- Huawei

- Nokia
- Nokia Solutions and Networks

Mobile Network Operators

- Deutsche Telekom
- Orange
- Telecom Italia
- Telefonica
- NTT DoCoMo

Academic and Research Organizations

- Aalborg University. Denmark
- Aalto University, Finland
- Chalmers Tekniska Hoegskola, Sweden
- Fraunhofer, Germany
- Institut Mines-Telecom, France
- Kungliga Tekniska Hoegskolan, Sweden
- National and Kapodistrian University of Athens, Greece
- Oulu University, Finland
- Poznan University of Technology, Poland
- Rheinisch-Westfaelische Technische Hochschule Aachen, Germany
- Technische Universitaet Kaiserslautern, Germany
- Universitaet Bremen, Germany
- Universidad Politecnica de Valencia, Spain

Automotive Industry

- BMW

So far (October 2013), METIS has published six deliverables on its website [9]. In addition to top-level system requirements for 5G, it has also published an interesting initial review of radio spectrum that could be

12.4 METIS PROJECT

available for 5G and its suitability for the task. A very wide radio spectrum area was assessed, from 380 MHz to 275 GHz. When looking for new spectrum bands, METIS set a minimum requirement that it should be possible to identify continuous spectrum of several hundred megahertz below 40.5 GHz and at least 1 GHz above 40.5 GHz.

The radio frequencies below 5925 MHz were not discussed in depth in the initial study because those frequencies will be considered in the WRC-15 preparation process. In any case, that spectrum is already heavily used and it will be difficult to allocate large bandwidths for 5G usage from this frequency band.

Table 12.3 contains the spectrum bands that were identified as potential bands for 5G [10].

The deadline for the current METIS project is in the end of April 2015. However, it is quite clear that there will be a continuation project of some form building on the results of this project. Also, most probably the results from METIS and its followup projects will be fed into future 3GPP LTE-A releases via company contributions (note that all nonacademic METIS members are also 3GPP member companies).

TABLE 12.3 POTENTIAL 5G FREQUENCY BANDS AS IDENTIFIED BY METIS

Band (GHz)	Size [GHz]	Priority
9.9–10.6	0.7	Medium-High
17.1–17.3	0.2	Low
17.7–19.7	2.0	Low
21.2–21.4	0.2	Low
27.5–29.5	2.0	Medium
31.0–31.3	0.3	Medium
31.8–33.4	1.6	High
36.0–37.0	1.0	Low
40.5–42.5	2	Medium
42.5–43.5	1	High
43.5–45.5	2	Low
45.5–47.0	1.5	High
47.2–50.2	3	High
50.4–52.6	2.2	Medium-Low
55.78–57.0	1.22	High
57–66	7	High
66–71	5	High
71–76	5	High
81–86	5	High

References

[1] 3GPP TR 22.803, v 12.2.0, Feasibility Study for Proximity Services (ProSe); 06/2013.

[2] 3GPP TR 23.703, v 0.7.1, Study on Architecture Enhancements to Support Proximity-Based Services (ProSe); 09/2013.

[3] 3GPP TS 22.368, v 12.2.0, Service Requirements for Machine-Type Communications (MTC); Stage 1; 03/2013.

[4] 3GPP TR 36.836, v 2.0.2, Study on Mobile Relay for Evolved Universal Terrestrial Radio Access (E-UTRA; 07/2013.

[5] Nakamura T., Nagata S., Benjebbour A., Kishiyama Y., Hai T., et al., "Trends in Small Cell Enhancements in LTE Advanced," *IEEE Communications*, February 2013, pp. 98–105.

[6] Larsson E. G., Tufvesson F., Edfors O., and Marzetta T. L., "Massive MIMO for Next Generation Wireless Systems," arXiv:1304.6690, 2013.

[7] Metis2020; *The EU initiative METIS paves the way for the mobile and wireless communications system for 2020 and beyond;* https://www.metis2020.com/press-events/press/20121127-metis-paves-the-way; last accessed January 1, 2014.

[8] Metis2020; The 5G future scenarios identified by METIS-the first step towards a 5G mobile and wireless communications system; https://www.metis2020.com/press-events/press/the-5g-future-scenarios-identified-by-metis; last accessed January 2, 2014.

[9] METIS website, https://www.metis2020.com; last accessed January 2, 2014.

[10] METIS project, Deliverable D5.1, "Intermediate Description of the Spectrum Needs and Usage Principles," August 30, 2013, www.metis2020.com/wp-content/uploads/deliverables/METIS_D5.1_v1.pdf; last accessed January 2, 2014.

Appendix: LTE and LTE-A Specifications

TS 36.101 User equipment (UE) radio transmission and reception

TS 36.104 Base station (BS) radio transmission and reception

TS 36.106 FDD repeater radio transmission and reception

TS 36.111 Location measurement unit (LMU) performance specification; network-based positioning systems in evolved universal terrestrial radio access network (E-UTRAN)

TS 36.112 Location measurement unit (LMU) conformance specification; network-based positioning systems in evolved universal terrestrial radio access network (E-UTRAN)

TS 36.113 Base station (BS) and repeater electromagnetic compatibility (EMC)

TS 36.116 Relay radio transmission and reception

TS 36.117 Relay conformance testing

TS 36.124 Electromagnetic compatibility (EMC) requirements for mobile terminals and ancillary equipment

TS 36.133 Requirements for support of radio resource management

TS 36.141 Base station (BS) conformance testing

TS 36.143 FDD repeater conformance testing

TS 36.171 Requirements for support of assisted global navigation satellite system (A-GNSS)

TS 36.201 LTE physical layer; general description

TS 36.211 Physical channels and modulation

TS 36.212 Multiplexing and channel coding

TS 36.213 Physical layer procedures

TS 36.214 Physical layer; measurements

TS 36.216 Physical layer for relaying operation

TS 36.300 Overall description; stage 2

TS 36.302 Services provided by the physical layer

TS 36.304 User equipment (UE) procedures in idle mode

TS 36.305 Stage 2 functional specification of user equipment (UE) positioning in E-UTRAN

TS 36.306 User Equipment (UE) radio access capabilities

TS 36.307 Requirements on user equipments (UEs) supporting a release-independent frequency band

TS 36.314 Layer 2—measurements

TS 36.321 Medium access control (MAC) protocol specification

TS 36.322 Radio link control (RLC) protocol specification

TS 36.323 Packet data convergence protocol (PDCP) specification

TS 36.331 Radio resource control (RRC); protocol specification

TS 36.355 LTE positioning protocol (LPP)

TS 36.401 Architecture description

TS 36.410 S1 general aspects and principles

TS 36.411 S1 layer 1

TS 36.412 S1 signaling transport

TS 36.413 S1 application protocol (S1AP)

TS 36.414 S1 data transport

TS 36.416 Mobile relay for E-UTRA SPEC WITHDRAWN

TS 36.420 X2 general aspects and principles

TS 36.421 X2 layer 1

TS 36.422 X2 signaling transport

TS 36.423 X2 application protocol (X2AP)

TS 36.424 X2 data transport

TS 36.440 General aspects and principles for interfaces supporting multimedia broadcast multicast service (MBMS) within E-UTRAN

TS 36.441 Layer 1 for interfaces supporting multimedia broadcast multicast service (MBMS) within E-UTRAN

TS 36.442 Signaling transport for interfaces supporting multimedia broadcast multicast service (MBMS) within E-UTRAN

TS 36.443 M2 application protocol (M2AP)

TS 36.444 M3 application protocol (M3AP)

TS 36.445 M1 data transport

TS 36.446 M1 user plane protocol SPEC WITHDRAWN

TS 36.455 LTE positioning protocol A (LPPa)

TS 36.456 SLm interface general aspects and principles

TS 36.457 SLm interface layer 1

TS 36.458 SLm interface signaling transport

TS 36.459 SLm interface application protocol (SLmAP)

TS 36.508 Common test environments for user equipment (UE) conformance testing

TS 36.509 Special conformance testing functions for user equipment (UE)

TS 36.521-1 User equipment (UE) conformance specification; radio transmission and reception; part 1: conformance testing

TS 36.521-2 User equipment (UE) conformance specification; radio transmission and reception; part 2: implementation conformance statement (ICS)

TS 36.521-3 User equipment (UE) conformance specification; radio transmission and reception; part 3: radio resource management (RRM) conformance testing

TS 36.523-1 User equipment (UE) conformance specification; part 1: protocol conformance specification

TS 36.523-2 User equipment (UE) conformance specification; part 2: implementation conformance statement (ICS) proforma specification

TS 36.523-3 User equipment (UE) conformance specification; part 3: test suites

TS 36.571-1 User equipment (UE) conformance specification; UE positioning in E-UTRA; part 1: minimum performance conformance SPEC WITHDRAWN

TS 36.571-2 User equipment (UE) conformance specification; UE positioning in E-UTRA; part 2: protocol conformance SPEC WITHDRAWN

TS 36.571-3 User Equipment (UE) conformance specification; UE positioning in E-UTRA; part 3: ICS SPEC WITHDRAWN

TS 36.571-4 User equipment (UE) conformance specification; UE positioning in E-UTRA; part 4: test suites SPEC WITHDRAWN

TS 36.571-5 User equipment (UE) conformance specification; UE positioning in E-UTRA; part 5: UE positioning test scenarios and assistance data SPEC WITHDRAWN

TR 36.800 Extended UMTS/LTE 800 work item technical report

TR 36.801 Measurement requirements SPEC WITHDRAWN

TR 36.803 User equipment (UE) radio transmission and reception SPEC WITHDRAWN

TR 36.804 Base station (BS) radio transmission and reception SPEC WITHDRAWN

TR 36.805 Study on minimization of drive-tests in next generation networks

TR 36.806 Relay architectures for E-UTRA (LTE-advanced)

TR 36.807 User equipment (UE) radio transmission and reception

TR 36.808 Carrier aggregation; base station (BS) radio transmission and reception

TR 36.809 Radio frequency (RF) pattern matching location method in LTE

TR 36.810 UMTS/LTE in 800 MHz for Europe

TR 36.811 Adding 2-GHz band LTE frequency division duplex (FDD) (band 23) for ancillary terrestrial component (ATC) of mobile satellite services (MSS) in North America

TR 36.812 LTE TDD 2600 MHz in US work item technical report

TR 36.813 Evolved universal terrestrial radio access (E-UTRA); LTE L-band technical report

TR 36.814 Further advancements for E-UTRA physical layer aspects

TR 36.815 Further advancements for E-UTRA; LTE-advanced feasibility studies in RAN WG4

TR 36.816 Study on signaling and procedure for interference avoidance for in-device coexistence

TR 36.817 Uplink multiple antenna transmission; base station (BS) radio transmission and reception

TR 36.818 Expanding 1900 MHz

TR 36.819 Coordinated multipoint operation for LTE physical layer aspects

TR 36.820 LTE for 700-MHz digital dividend

TR 36.821 Extended UMTS/LTE 1500 work item technical report

TR 36.822 LTE radio access network (RAN) enhancements for diverse data applications

TR 36.823 Carrier aggregation enhancements; user equipment (UE) and base station (BS) radio transmission and reception

TR 36.824 LTE coverage enhancements

TR 36.826 Relay radio transmission and reception

TR 36.827 LTE advanced carrier aggregation band 41 SPEC WITHDRAWN

TR 36.828 Further enhancements to LTE time division duplex (TDD) for downlink-uplink (DL-UL) interference management and traffic adaptation

TR 36.829 Enhanced performance requirement for LTE user equipment (UE)

TR 36.830 LTE-advanced carrier aggregation (CA) in band 38 SPEC WITHDRAWN

TR 36.831 LTE-advanced carrier aggregation (CA) in band 7 SPEC WITHDRAWN

TR 36.832 LTE in the 1670–1675 MHz band for the United States

TR 36.833 Intraband noncontiguous CA for band 4 for LTE SPEC WITHDRAWN

TR 36.833-1-03 LTE-advanced intraband contiguous carrier aggregation (CA) in band 3

TR 36.833-1-07 LTE-advanced intraband contiguous carrier aggregation (CA) in band 7

TR 36.833-1-23 LTE-advanced intraband contiguous carrier aggregation (CA) in band 23

TR 36.833-1-27 LTE-advanced intraband contiguous carrier aggregation (CA) in band 27

TR 36.833-1-38 LTE-advanced intraband contiguous carrier aggregation (CA) in band 38

TR 36.833-1-39 LTE-advanced intraband contiguous carrier aggregation (CA) in band 39

TR 36.833-1-41 LTE-advanced intraband contiguous carrier aggregation (CA) in band 41

TR 36.833-2-03 LTE-advanced intraband noncontiguous carrier aggregation (CA) in band 3

TR 36.833-2-04 LTE-advanced intraband noncontiguous carrier aggregation (CA) in band 4

TR 36.833-2-23 LTE-advanced intraband noncontiguous carrier aggregation (CA) in band 23

TR 36.833-2-25 LTE-advanced intraband noncontiguous carrier aggregation (CA) in band 25

TR 36.833-5-41 LTE-advanced intraband contiguous carrier aggregation (CA) in band 41 for 3 down-links (DL)

TR 36.834 LTE advanced intraband contiguous carrier aggregation in band 3 SPEC WITHDRAWN

TR 36.835 Intraband contiguous CA in band 1 SPEC WITHDRAWN

TR 36.836 Mobile relay for E-UTRA

TR 36.837 Public safety broadband high power user equipment (UE) for band 14

TR 36.838 Intraband contiguous CA in band 27 for LTE SPEC WITHDRAWN

TR 36.839 Evolved universal terrestrial radio access (E-UTRA); mobility enhancements in heterogeneous networks

TR 36.840 LTE 450 MHz in Brazil work item technical report

TR 36.841 Evolved universal terrestrial radio access (E-UTRA); intraband noncontiguous carrier aggregation (CA) for band 25 for LTE SPEC WITHDRAWN

TR 36.842 Study on small cell enhancements for E-UTRA and E-UTRAN—higher-layer aspects

TR 36.843 Feasibility study on LTE device-to-device proximity services—radio aspects

TR 36.844 Study on expansion of LTE_FDD_1670_US to include 1670–1680 MHz band for LTE in the US

TR 36.845 Intraband contiguous CA for band 39 for LTE SPEC WITHDRAWN

TR 36.846 LTE in the US wireless communications service (WCS) band

TR 36.847 LTE time division duplex (TDD)–frequency division duplex (FDD) joint operation including carrier aggregation (CA)

TR 36.848 Smart congestion mitigation in E-UTRAN

TR 36.850 Interband carrier aggregation

TR 36.851 Interband carrier aggregation

TR 36.853 Technical report for 3 band carrier aggregation with single uplink

TR 36.855 Feasibility of positioning enhancements for E-UTRA

TR 36.860 LTE-advanced dual uplink interband carrier aggregation (CA)

TR 36.861 Technical report for study item: study on LTE FDD in the bands 1980–2010 MHz and 2170–2200 MHz

TR 36.863 Feasibility of CRS interference mitigation for LTE homogenous deployment

TR 36.864 Intraband contiguous carrier aggregation in band 23 SPEC WITHDRAWN

TR 36.865 Intraband noncontiguous carrier aggregation in band 23 SPEC WITHDRAWN

TR 36.866 Network-assisted interference cancellation and suppression for LTE

TR 36.868 Study on evolved universal terrestrial radio access (E-UTRA)

TR 36.871 Downlink multiple input multiple output (MIMO) enhancement for LTE-advanced

TR 36.872 Small cell enhancements for E-UTRA and E-UTRAN—physical layer aspects

TR 36.873 3D-channel model for LTE

TR 36.874 Coordinated multipoint (COMP) operation for LTE with nonideal backhaul

TR 36.887 Study on energy saving enhancement for E-UTRAN

TR 36.888 Study on provision of low-cost machine-type communications (MTC) user equipments (UEs) based on LTE

TR 36.902 Self-configuring and self-optimizing network (SON) use cases and solutions

TR 36.903 Derivation of test tolerances for radio resource management (RRM) conformance tests

TR 36.912 Feasibility study for further advancements for E-UTRA (LTE-advanced)

TR 36.913 Requirements for further advancements for evolved universal terrestrial radio access (E-UTRA) (LTE-advanced)

TR 36.921 FDD home eNode B (HeNB) radio frequency (RF) requirements analysis

TR 36.922 TDD home eNode B (HeNB) radio frequency (RF) requirements analysis

TR 36.927 Potential solutions for energy saving for E-UTRAN

TR 36.931 Radio frequency (RF) requirements for LTE pico node B

TR 36.932 Scenarios and requirements for small cell enhancements for E-UTRA and E-UTRAN

TR 36.938 Improved network controlled mobility between E-UTRAN and 3GPP2/mobile WiMAX radio technologies

TR 36.942 Radio frequency (RF) system scenarios

TR 36.956 Repeater planning guidelines and system analysis SPEC WITHDRAWN

About the Author

Juha Korhonen was born in Suonenjoki, Finland. He holds two university degrees in telecommunications from Lappeenranta University of Technology, Finland, and a Ph.D. from Cambridge University, UK.

He joined the Nokia Research Center in Espoo near Helsinki in 1988. At that time, his work involved the development of protocol software testers and tools. From 1993 to 1995, he worked as a design engineer at Nokia Mobile Phones, Camberley, UK, developing JDC mobile phones for the Japanese market. In 1995, he returned to Nokia Research Center, this time to work on advanced DECT research. During 1996 and 1997 he was the leader of the protocol stack team that developed the WCDMA prototype network.

In 1997 he moved to Cambridge, UK, to join TTPCom, Ltd. He was first involved in a satellite phone development program, and after 1999 he worked in various research projects related to 3G technologies. Between 2006 and 2009 he worked as an independent telecoms consultant.

In 2009 he joined the European Telecommunications Standards Institute in Sophia Antipolis, France, where he works as a project manager for 3GPP projects; currently, he is responsible for the 3GPP RAN3 working group.

He lives in the little village of Saint Cézaire in Southern France with his wife, Akiko, and baby daughter, Koyuki.

Index

1x advanced, 25
1xEV-DO, 23–24
1xRTT, 23
3GPP
    introduction to, 18–19
    organizational partners, 196
    releases, 20–22
    releases timeline, 22
    specification series, 210
    structure. *see* internal structure
    UMTS, 19–20
    See also Third generation mobile telecommunications
3GPP2
    EV-DO advanced, 24–25
    evolution data and voice (1xEVDV), 26
    introduction to, 22–23
    1x advanced, 25
    1xEV-DO, 23–24
    1xRTT, 23
    UMB, 26
    See also Third generation mobile telecommunications
64QAM, 74

A

Access stratum (AS), 89
Adaptive frequency reuse, 24–25
Adaptive modulation and coding (AMC), 97
Advanced mobile phone service (AMPS), 6–7
Air interface
    5G, 258–59
    channel concepts, 60
    frame structure, 61
    OFDM, 31–33
    physical layer, 59–86
    protocol stack, 60, 89–123
    resource scheduling, 55
Almost blank subframes (ABS), 43, 233
Amplitude shift keying (ASK), 71
Antenna ports, mapping to, 76
Antennas, 5G, 262–63
Automatic neighbor relations (ANR), 235

B

Backwards compatibility
    maintenance of, 211
    practical reasons for, 212
    specifications under work and, 212
Bandwidth, 5G, 259–61
Beamforming, 263
Binary phase shift keying (BPSK), 72, 73
Broadcast channel (BCH), 96–97
Broadcast control channel (BCCH), 103

C

Carrier aggregation, 226–29
    component carriers, 227
    defined, 226
    directional macro cells with, 254
    enhancements, 45
    in HetNets, 256
    interband noncontiguous, 227, 228, 230–31
    intraband contiguous, 227, 229
    intraband noncontiguous, 227, 228
    introduction of, 31, 42
    methods, 227
    in multisite HetNet, 228
CDMA2000, 18, 22–23
CdmaOne, 15

Cell search
    defined, 169
    PCI, 171
    PSS sequences, 169–70
    SSS sequence, 170
    See also Procedures
Cell-specific reference signals (CRS), 81, 82
Cell-specific reference symbol (CRS), 233
Cellular systems, 5, 6
Channel coding
    convolutional codes, 66
    summary, 68
    turbo codes, 65–66
Channel interleaving, 69–70
Channel state information (CSI)
    downlink MIMO and, 222
    reference signals, 81, 84
Ciphering, 132
Circuit switched data (CSD), 11
Circular buffer, 70, 71
Closed subscriber group (CSG), 136–37, 186
C-NETZ, 8
Codebooks, 76
Code division multiple access (CDMA)
        standard IS-95
    defined, 14–15
    TDMA versus, 15
Common control channel (CCCH), 103
Connection control, RRC, 114–16
Connection mobility control (CMC), 130
Control plane protocol stack, 91
Convolutional codes, 66
Coordinated scheduling and beamforming
        (CS/CB), 237–38
Coordinate multipoint transmission and
        reception (CoMP)
    defined, 44, 236
    downlink, 237–38
    overview of, 236–37
    uplink, 238
Core network, EPC, 147–65
CSG inbound HO
    defined, 186
    illustrated, 187
    procedure, 186–89
    See also Procedures
Cyclic redundancy check (CRC)
    defined, 63
    insertion, 63–64

D

D-AMPS, 13–14
Dedicated bearer activation
    defined, 191
    illustrated, 192
    procedure, 191–93
Dedicated control channel (DCCH), 103–4
Dedicated traffic channel (DTCH), 104
Development cycle, 209–10
Discontinuous reception (DRX), 97
Distributed network scheduler, 25
Donor eNodeB (DeNB), 142–45
Downlink CoMP, 237–38
Downlink control information (DCI), 95
Downlink MIMO, 221–24
Downlink shared channel (DL-SCH), 97
Downlink signals, 81–85
Downlink transport channels, 96–98
Dynamic resource allocation (DRA), 130

E

Energy saving
    capacity cells on/off, 216
    HetNets, 257
    by increasing cell size, 218
    overview of, 215
    predefined configurations, 217
    problems to solve, 216–17
    summary, 219
    UE measurement method, 218
Enhanced connection management, 25
Enhanced data rates for global evolution
        (EDGE), 12–13
Enhanced intercell interference coordination
        (eICIC), 229–33
    defined, 229
    interference and, 43
    resource allocation, 232

ENodeBs, 127–34
    access stratum security, 131–32
    connection mobility control (CMC), 130
    donor (DeNB), 142–45
    dynamic resource allocation (DRA), 130
    E-UTRAN architecture, 125–26
    functionality, 128–34
    home (HeNB), 134–42
    intercell interference coordination (ICIC), 130–31
    introduction to, 127–28
    IP header compression and encryption of user data, 132
    load balancing (LB), 131
    measurement and measurement reporting configuration, 131
    protocol stacks, 127–28
    radio admission control (RAC), 130
    radio bearer control (RBC), 130
    radio resource management (RRM), 129
    routing of user plane data, 133
    scheduling and transmission of broadcast information, 133
    scheduling and transmission of paging messages, 133
    scheduling and transmission of PWS messages, 133–34
    selection of MME, 133
European Telecommunications Standards Institute (ETSI), 17
E-UTRAN architecture
    defined, 126
    eNodeBs, 125–26
    illustrated, 126
    interfaces and protocol tasks, 212–14
    relay nodes (RN), 126–27
    specifications, 212–14
EV-DO advanced, 24–25
Evolution data and voice (1xEVDV), 26
Evolved multimedia broadcast multicast service (eMBMS)
    counting function, 233–34
    defined, 233
    high-level architecture, 234
Evolved packet core (EPC)
    architecture, 148–55
    control plane, 155–60
    defined, 33
    evolved serving mobile location center (E-SMLC), 151–54
    functionality, 147
    gateway mobile location center (GMLC), 155
    home subscriber server (HSS), 151, 152–53
    interfaces and protocols, 155–65
    interface summary, 162–65
    introduction to, 147–48
    mobile management entity (MME), 148–50
    packet data network gateway (PDN GW), 150–51
    policy and charging rules function (PCRF), 155
    S1-MME interface, 156
    S1-U interface, 161–62
    S4 interface, 162, 163
    S5/S8 interface, 158
    S6a interface, 159–60
    S7a interface, 160, 161
    S10 interface, 158, 159
    S11 interface, 158–59
    S12 interface, 162, 163
    S13 interface, 160, 161
    serving GW, 150
    UE-MME interface, 156–57
    user plane, 160–62
Evolved serving mobile location center (E-SMLC)
    defined, 151
    MME connection, 154
    positioning data, 154
    positioning methods, 153–54
Evolving LTE-A
    heterogeneous networks (HetNets), 255–57
    machine-type communications (MTC), 248–51
    mobile relays, 251–55
    proximity services, 244–48
Extended TACS (ETACS), 7

F

Fast Fourier Transform (FFT), 52
Fifth generation (5G)
    air interface, 258–59
    bandwidth, 259–61

Fifth generation (5G) (continued)
  caution with terminology, 257
  improvements, 258
  introduction to, 257–58
  M2M communications support, 263–64
  METIS-identified frequency bands, 267
  multiple antennas, 262–63
  network architecture, 261–62
  spectrum use, 260
Forward error correction (FEC) schemes, 64
Frequency division duplex (FDD)
  defined, 19
  OFDM and, 52
  radio frame structure, 53, 61
Frequency selective fading, 48
Frequency shift keying (FSK), 71, 72
Further enhanced intercell interference coordination (FeICIC), 45
Future developments
  evolving LTE-A, 244–57
  fifth generation, 257–65
  heterogeneous networks (HetNets), 255–57
  introduction to, 243
  machine-type communications (MTC), 248–51
  METIS project, 265–67
  mobile relays, 251–55
Future-proof, 40

G

Gateway mobile location center (GMLC), 155
Gaussian minimum shift keying (GMSK), 72
General packet radio service (GPRS), 11–12
Global system for mobile communication (GSM), 10–13
Guard intervals, 50–51

H

Handover
  defined, 182
  illustrated, 184
  phases, 183
  procedure, 183–85
  See also Procedures

Heterogeneous networks (HetNets)
  carrier aggregation, 256
  defined, 255
  illustrated, 256
  network energy savings, 257
High-speed downlink packet access (HSDPA), 21
High-speed uplink packet access (HSUPA), 21
Home eNodeB (HeNB), 134–42
  access modes, 134–35
  architecture, 135
  closed subscriber group (CSG), 136–37
  defined, 134
  gateway, 137
  gateway control plane protocol stacks, 138
  gateway user plane control stacks, 138
  interference solutions, 136
  introduction to, 134–36
  location ability, 136
  mobility, 137
  service requirements, 136
  traffic offloading, 137–42
Home subscriber server (HSS), 151, 152–53
Hybrid-ARQ (HARQ) processing, 68–69, 97

I

Inband relays, 225–26
Initial context setup
  defined, 179
  illustrated, 180
  procedure, 179–82
Integrated dispatch enhanced network (iDEN), 16
Interband noncontiguous carrier aggregation, 227, 228
Intercarrier interference (ICI)
  defined, 49
  illustrated, 50
  suppression, 51
Intercell interference coordination (ICIC), 130–31
Interference over thermal (IoT) measurements, 218
Internal structure
  illustrated, 196
  mobile competence center (MCC), 202
  overview of, 195–97
  TSG CT, 200–201

TSG GERAN, 201–2
TSG RAN, 197–99
TSG SA, 199–200
Intraband contiguous carrier aggregation, 227, 229
Intraband noncontiguous carrier aggregation, 227, 228, 230–31
Inverse-FFT (IFFT), 52
IP multimedia subsystem (IMS), 238–39
IS-95 CDMA, 14–15

J

Japanese Total Access Communication System (JTACS), 8

L

L3 relays. *See* Relays
Layer mapping and precoding, 75
LIPA, 139, 140
Load balancing (LB), 131
Location areas (LAs), 239–40
Logical channels
    broadcast control channel (BCCH), 103
    common control channel (CCCH), 103
    control, 103–4
    dedicated control channel (DCCH), 103–4
    dedicated traffic channel (DTCH), 104
    defined, 102
    multicast control channel (MCCH), 103
    multicast traffic channel (MTCH), 104
    paging control channel (PCCH), 103
    traffic, 104–5
Long-term evolution (LTE)
    advanced. *see* LTE-A
    architecture and migration, 36
    architecture and migration requirements, 39
    codebooks, 76
    co-existence and interworking, 35–36
    complexity, 36, 39
    control-plane capacity, 34, 38
    control-plane latency, 34
    coverage, 35, 38
    defined, 1
    enhanced MBMS, 35, 38–39
    latency improvement, 37
    mobility, 35, 38
    OFDM air interface, 33
    OFDM symbols, 53
    peak data rate, 34
    radio resource management requirements, 36
    RAN, 30–31
    relay architecture, 225
    release schedule, 206
    requirements, 33–40
    specifications, 1, 269–75
    spectrum efficiency, 34–35, 38
    spectrum flexibility, 35, 39
    uplink, 56–57
    user-plane latency, 34, 38
    user throughput, 34, 38
    voice and, 238–41
LTE-A
    3GPP characterization of, 41
    average spectrum efficiency, 42
    carrier aggregation, 31, 42, 45, 226–29
    coordinate multipoint transmission and reception (CoMP), 44, 236–38
    defined, 26, 40
    downlink physical channels, 77–79
    energy saving, 215–19
    enhanced intercell interference coordination (eICIC), 43, 229–33
    evolved multimedia broadcast multicast service (eMBMS), 233–34
    features, 215–41
    further enhanced intercell interference coordination (FeICIC), 45
    higher order MIMO, 42, 43
    high-level requirements, 40
    introduction to, 29
    MBMS service continuity, 45
    MIMO, 219–24
    network architecture illustration, 30
    new features, 42
    physical channels, 76–81
    proximity services, 244–48
    relays, 43, 224–26

284 INDEX

LTE-A (continued)
    Release 11 enhancements/new features, 44–45
    self-organizing networks (SON), 44, 235–36
    specifications, 269–75
    system architecture, 29–30
    uplink physical channels, 80–81
    voice and, 238–41

M

Machine-to-machine (M2M) communication, 263–64
Machine-type communications (MTC)
    applications, 249–51
    defined, 248
    service areas, 249
Market representation partners (MRP), 197
Master information block (MIB), 96, 110–11
Maximum likelihood sequence estimator (MLSE), 67
MBSFN reference signals, 81, 83
Medium access control (MAC)
    defined, 90
    functionality in UE/eNodeB, 94
    functions supported by, 92
    general, 90–96
    logical channel prioritization, 95
    relay nodes (RN), 92
    structure illustration, 93
    transport channels, 96–98
    UE, 92, 93
    See also Protocol stack
METIS project
    deadline, 267
    defined, 265
    members, 265–66
    potential 5G frequency bands, 267
    published deliverables, 266–67
    requirements, 265
MIMO, 219–24
    channel state information (CSI) and, 222
    defined, 219
    downlink, 221–24
    general system, 220
    higher order, 42, 43
    MU-MIMO, 221–22

    overview, 219–20
    spatial multiplexing, 219, 224
    SU-MIMO, 221–22
    transmission modes and, 223
    uplink, 224
Minimization drive tests (MDT), 121–22, 236
Minimum shift keying (MSK), 72
Mobile competence center (MCC), 202
Mobile management entity (MME), 148–50
Mobile relays
    advantages/disadvantages of, 253
    cost effect, 253
    functions of, 253
    high-speed train challenges, 251–52
    latency and, 254
    ownership, 254
    RRUs and, 255
    tunnels and, 252
Mobile telecommunications
    3GPP, 17–22
    3GPP2, 22–26
    AMPS, 6–7
    C-NETZ, 8
    D-AMPS, 13–14
    generations and technologies, 27
    GSM, 10–13
    history of, 5–28
    iDEN, 16
    introduction to, 5–6
    IS-95 CDMA, 14–15
    Japanese systems, 8
    NMT, 7–8
    PDC, 16
    Radiocom2000, 8
    second generation, 9–17
    TACS, 7
    TETRA, 17
    third generation, 17–26
Mobility load balancing (MLB), 235
Mobility robustness optimization (MRO), 235–36
Modulation, 71–74
Modulation and coding scheme (MCS), 95
Multicast channel (MCH), 98
Multicast control channel (MCCH), 103

INDEX

Multicast traffic channel (MTCH), 104
Multimedia broadcast multicast services (MBMS)
    as broadcast/multicast service, 44
    enhanced, 35, 38–39, 44
    network capacity consumption, 21
    service continuity, 45
Multiple input, multiple output. *See* MIMO
Multiplexing, 94
MU-MIMO, 221–22

N

Narrowband advanced mobile phone service (NAMPS), 6
Narrowband TACS, 8
Network architecture, 5G, 261–62
Network load balancing (NLB), 24
NodeB's, 29
Nonaccess stratum (NAS), 89, 157
Nordic mobile telephone (NMT), 7–8
NTT DoCoMo, 8
Numbering, specification, 210–11

O

Observed time difference of arrival (OTDOA), 83–84
OFDM
    advantages over WCDMA, 31–32
    air interface, 31–33
    defined, 47
    frequency selective fading, 48
    guard intervals, 50–51
    intercarrier interference (ICI) and, 49, 50, 51
    introduction to, 47–51
    in LTE downlink and uplink, 52
    LTE symbols, 53
    narrowband carriers, 31, 47–48
    number of subcarriers, 54
    orthogonal subcarriers, 49
    principles, 51–56
    resource block, 54–55
    signal processing chain, 52
    subcarrier spacing, 48
    summary, 57–58

Orthogonal frequency division multiplexing. *See* OFDM
Outband relays, 226
Overview, this book, 2–3

P

Packet data convergence protocol (PDCP)
    defined, 105
    handover support, 106
    layer, control plane, 107
    layer, user plane, 106
    lossless handover, 107–8
    security, 106
    tasks, 105
    See also Protocol stack
Packet data network gateway (PDN GW), 150–51
Paging channel (PCH), 97–98
Paging control channel (PCCH), 103
Parallel concatenated convolutional code (PCCC), 65
Personal digital cellular (PDC), 16
Phantom cells, 262
Phase shift keying (PSK), 71
Physical channels
    downlink, 77–79
    EPDCCH, 78
    PBCH, 77
    PCFICH, 78
    PDCCH, 78
    PDSCH, 79
    PHICH, 79
    PMCH, 79
    PRACH, 81
    PUCCH, 80
    PUSCH, 80
    R-PDCCH, 79
    uplink, 80–81
Physical downlink shared channel (PDSCH), 63–64
Physical layer
    channel coding, 64–68
    channel interleaving, 69–70
    CRC insertion, 63–64
    downlink data path, 63
    downlink signals, 81–85
    general concepts, 60–62

286 INDEX

Physical layer (continued)
   hybrid-ARQ processing, 68–69
   introduction to, 59–60
   processing, 62
   rate matching, 69–70
   reference signals, 81–84
   signals, 81–86
   synchronization signals, 85
   uplink signals, 85–86
Physical layers
   layer mapping and precoding, 75–76
   mapping to assigned resources and antenna ports, 76
   modulation, 71–74
   scrambling, 70–71
Policy and charging rules function (PCRF), 155, 191
Position reference signals (PRS), 81, 83–84
Precoding, 75–76
Primary synchronization signal (PSS), 84
Procedures, 169–93
   cell search, 169–71
   CSG inbound HO, 186–89
   dedicated bearer activation, 191–93
   handover (X2 interface), 182–86
   initial context setup, 179–82
   introduction to, 169
   random access, 171–73
   S1 release, 189–91
   tracking area update, 174–79
Project Coordination Group (PCG), 196–97
Protocol stack, 89–123
   eNodeB, 127–28
   medium access control (MAC), 90–98
   packet data convergence protocol (PDCP), 105–8
   radio link control (RLC), 99–105
   radio resource control (RRC), 108–23
Proximity services
   business case for, 247–48
   closed user group, 247
   concept, 244
   defined, 244
   emergency services radio resource controller, 247

introduction of, 247
   scenarios, 245–46
Pruning, 69

Q

Quaternary phase shift keying (QPSK), 72, 73

R

Radio access network (RAN), 125–45
   eNodeB, 127–34
   E-UTRAN architecture, 125–27
   home eNodeB, 134–42
   introduction to, 125
   relay nodes (RN), 142–45
Radio access technology (RAT), 35, 39
Radio admission control (RAC), 130
Radio bearer control (RBC), 130
Radiocom2000, 8
Radio link control (RLC)
   acknowledged mode entity in, 102
   control channels, 103–4
   defined, 99
   functions, 101–5
   general, 99–101
   logical channels, 102–5
   traffic channels, 104–5
   transport mode entity, 10
   unacknowledged mode entity, 101
   See also Protocol stack
Radio network controllers (RNC), 29
Radio resource control (RRC)
   cell types, 118
   connection control, 114–16
   connection establishment, 115
   connection reconfiguration, 115
   connection release procedure, 116
   as control plane entity, 89
   counter check procedure, 115–16
   defined, 108
   emergency services, 247
   event-triggered criteria, 119
   functions, 110–23

generic protocol error handling, 120–21
integrity protection algorithm, 115
inter-RAT mobility, 116–17
introduction to, 108
master information block (MIB), 110–11
measurement configuration and reporting, 117–20
measurement logging and reporting support, 121–22
measurement parameters, 118
measurement rules, 118–19
measurement types, 117–18
proximity indication, 116
security management, 115
self-configuration/self-optimization support, 121
system information blocks (SIBs), 110–14
UE states, 108–10
See also Protocol stack
Radio resource management (RRM), 129
Radio telephony systems, 5, 6
Random access channel (RACH), 98
Random access procedure
    contention based, 172
    defined, 171
    response timing, 173
    steps, 172–73
    uplink time synchronization and, 171–72
Rate matching
    circular buffer, 70, 71
    defined, 69
    HARQ functionality, 70
    illustrated, 69
Reference signals, 81–84
Relay nodes (RN), 126–27
    defined, 142
    inband and outband types, 142
    network architecture, 143
Relays
    configurations, 225–26
    defined, 224
    inband, 225–26
    LTE architecture, 225
    mobile, 251–55
    need for, 43
    outband, 226
    reasons for use, 224–25
Releases
    defined, 205
    LTE schedule, 206
    Release 8, 206–7
    Release 9, 207
    Release 10, 207–8
    Release 11, 208
    Release 12, 208–9
    Robust header compression (ROHC) protocol, 132

S

S1 release procedure
    defined, 189
    illustrated, 190
    steps, 189–90
SC-FDMA
    advantages, 57
    defined, 56
    receiver block diagram, 57
    signal processing chain, 56
Scrambling, 70–71
Secondary synchronization signal (SSS), 84
Second generation mobile telecommunications, 9–17
Selected IP traffic offload. *See* SIPTO
Self-organizing networks (SON)
    ANR, 235
    coordination between functions, 236
    coverage and capacity optimization, 236
    defined, 235
    enhancements, 44, 45
    improvements, 235–36
    MLB, 235
    MRO, 235–36
    RACH optimization function, 236
    RRC and, 121
Serving gateway, 150
Single-carrier FDMA. *See* SC-FDMA
Single carrier multilink, 25

SIPTO
    above RAN, 139
    architecture illustrations, 139–42
    defined, 139
    at local network (L-GW collocated with (H) eNB), 139–41
    at local network (standalone GW), 141–42
Smart carrier management, 25
SMS-over-SG interface, 240, 241
Soft input soft output (SISO) decoders, 65
Sounding reference signals (SRS), 85
Spatial multiplexing, 219, 224
Specifications, 195–214
    backwards compatibility, 211–12
    E-UTRAN, 212–14
    internal structure, 195–202
    introduction to, 195
    LTE and LTE-A, 269–75
    numbering, 210–11
    standardization process, 202–10
    technical, 211
Standardization process
    development cycle, 209–10
    introduction to, 202
    releases, 205–9
    version numbering, 204–5
    work items, 202
Stratum model, 90
SU-MIMO, 221–22
Synchronization signals, 85
System architecture evolution (SAE), 33
System information blocks (SIBs), 110–14

T

Technical specification groups (TSG)
    defined, 197
    TSG CT, 200–201
    TSG GERAN, 201–2
    TSG RAN, 197–99
    TSG SA, 199–200
TETRA, 17
Third generation mobile telecommunications

3GPP, 18–22
3GPP2, 22–26
introduction to, 17–18
Time division duplex (TDD)
    defined, 19
    OFDM and, 52
    radio frame structure, 62
Time division multiple access (TDMA) systems, 10
Total access communication system (TACS), 7
Tracking areas (TAs), 174, 175, 239–40
Tracking area updates
    basic, 176
    with MME and S-GW changes, 177
    periodic, 175
    procedure, 174–79
    steps, 175–76, 178–79
    tracking area identity (TAI) list, 174, 175
    See also Procedures
Traffic offloading
    defined, 137–38
    LIPA, 139, 140
    SIPTO, 139–42
    See also Home eNodeB (HeNB)
Transmission modes, downlink MIMO, 223
Transport channels
    broadcast channel (BCH), 96–97
    defined, 96
    downlink, 96–98
    downlink shared channel (DL-SCH), 97
    multicast channel (MCH), 98
    paging channel (PCH), 97–98
    random access channel (RACH), 98
    uplink, 98
    uplink shared channel (UL-SCH), 98
Transport format, 95
Turbo codes, 65–66

U

UE-specific reference signals (DM-RS), 81, 83
Ultra mobile broadband (UMB), 26
Universal mobile telecommunications system (UMTS), 18, 19–20

Uplink CoMP, 238
Uplink MIMO, 224
Uplink shared channel (UL-SCH), 98
Uplink signals, 85–86
Uplink transport channels, 98
User plane protocol stack, 91
UTRAN system architecture, 30

V

Version numbering, 204–5
Voice, LTE and, 238–41

W

Wavelengths and frequencies, 261
Wideband code division multiple access (WCDMA), 19–20
  defined, 18
  interference-limited cells, 32
  OFDM advantages over, 31–32
Work items, in standardization process, 202–4
World Radio Conference 15 (WRC-15), 261

Recent Titles in the Artech House Mobile Communications Series

John Walker, Series Editor

3G CDMA2000 Wireless System Engineering, Samuel C. Yang

3G Multimedia Network Services, Accounting, and User Profiles, Freddy Ghys, Marcel Mampaey, Michel Smouts, and Arto Vaaraniemi

802.11 WLANs and IP Networking: Security, QoS, and Mobility, Anand R. Prasad and Neeli R. Prasad

Achieving Interoperability in Critical IT and Communications Systems, Robert I. Desourdis, Peter J. Rosamilia, Christopher P. Jacobson, James E. Sinclair, and James R. McClure

Advances in 3G Enhanced Technologies for Wireless Communications, Jiangzhou Wang and Tung-Sang Ng, editors

Advances in Mobile Information Systems, John Walker, editor

Advances in Mobile Radio Access Networks, Y. Jay Guo

Applied Satellite Navigation Using GPS, GALILEO, and Augmentation Systems, Ramjee Prasad and Marina Ruggieri

Artificial Intelligence in Wireless Communications, Thomas W. Rondeau and Charles W. Bostian

Broadband Wireless Access and Local Network: Mobile WiMax and WiFi, Byeong Gi Lee and Sunghyun Choi

CDMA for Wireless Personal Communications, Ramjee Prasad

CDMA Mobile Radio Design, John B. Groe and Lawrence E. Larson

CDMA RF System Engineering, Samuel C. Yang

CDMA Systems Capacity Engineering, Kiseon Kim and Insoo Koo

CDMA Systems Engineering Handbook, Jhong S. Lee and Leonard E. Miller

Cell Planning for Wireless Communications, Manuel F. Cátedra and Jesús Pérez-Arriaga

Cellular Communications: Worldwide Market Development, Garry A. Garrard

Cellular Mobile Systems Engineering, Saleh Faruque

Cognitive Radio Techniques: Spectrum Sensing, Interference Mitigation, and Localization, Kandeepan Sithamparanathan and Andrea Giorgetti

The Complete Wireless Communications Professional: A Guide for Engineers and Managers, William Webb

Digital Communication Systems Engineering with Software-Defined Radio, Di Pu and Alexander M. Wyglinski

EDGE for Mobile Internet, Emmanuel Seurre, Patrick Savelli, and Pierre-Jean Pietri

Emerging Public Safety Wireless Communication Systems, Robert I. Desourdis, Jr., et al.

The Future of Wireless Communications, William Webb

Geographic Information Systems Demystified, Stephen R. Galati

GPRS for Mobile Internet, Emmanuel Seurre, Patrick Savelli, and Pierre-Jean Pietri

GPRS: Gateway to Third Generation Mobile Networks, Gunnar Heine and Holger Sagkob

GSM and Personal Communications Handbook, Siegmund M. Redl, Matthias K. Weber, and Malcolm W. Oliphant

GSM Networks: Protocols, Terminology, and Implementation, Gunnar Heine

GSM System Engineering, Asha Mehrotra

Handbook of Land-Mobile Radio System Coverage, Garry C. Hess

Handbook of Mobile Radio Networks, Sami Tabbane

High-Speed Wireless ATM and LANs, Benny Bing

Inside Bluetooth Low Energy, Naresh Gupta

Interference Analysis and Reduction for Wireless Systems, Peter Stavroulakis

Introduction to 3G Mobile Communications, Second Edition, Juha Korhonen

Introduction to 4G Mobile Communications, Juha Korhonen

Introduction to Communication Systems Simulation, Maurice Schiff

Introduction to Digital Professional Mobile Radio, Hans-Peter A. Ketterling

Introduction to GPS: The Global Positioning System, Ahmed El-Rabbany

An Introduction to GSM, Siegmund M. Redl, Matthias K. Weber, and Malcolm W. Oliphant

Introduction to Mobile Communications Engineering, José M. Hernando and F. Pérez-Fontán

Introduction to Radio Propagation for Fixed and Mobile Communications, John Doble

Introduction to Wireless Local Loop, Broadband and Narrowband, Systems, Second Edition, William Webb

IS-136 TDMA Technology, Economics, and Services, Lawrence Harte, Adrian Smith, and Charles A. Jacobs

Location Management and Routing in Mobile Wireless Networks, Amitava Mukherjee, Somprakash Bandyopadhyay, and Debashis Saha

LTE Air Interface Protocols, Mohammad T. Kawser

Metro Ethernet Services for LTE Backhaul, Roman Krzanowski

Mobile Data Communications Systems, Peter Wong and David Britland

Mobile IP Technology for M-Business, Mark Norris

Mobile Satellite Communications, Shingo Ohmori, Hiromitsu Wakana, and Seiichiro Kawase

Mobile Telecommunications Standards: GSM, UMTS, TETRA, and ERMES, Rudi Bekkers

Mobile-to-Mobile Wireless Channels, Alenka Zajić

Mobile Telecommunications: Standards, Regulation, and Applications, Rudi Bekkers and Jan Smits

Multiantenna Digital Radio Transmission, Massimiliano Martone

Multiantenna Wireless Communications Systems, Sergio Barbarossa

Multi-Gigabit Microwave and Millimeter-Wave Wireless Communications, Jonathan Wells

Multipath Phenomena in Cellular Networks, Nathan Blaunstein and Jørgen Bach Andersen

Multiuser Detection in CDMA Mobile Terminals, Piero Castoldi

OFDMA for Broadband Wireless Access, Slawomir Pietrzyk

Personal Wireless Communication with DECT and PWT, John Phillips and Gerard MacNamee

Practical Wireless Data Modem Design, Jonathon Y. C. Cheah

Prime Codes with Applications to CDMA Optical and Wireless Networks, Guu-Chang Yang and Wing C. Kwong

Quantitative Analysis of Cognitive Radio and Network Performance, Preston Marshall

QoS in Integrated 3G Networks, Robert Lloyd-Evans

Radio Engineering for Wireless Communication and Sensor Applications, Antti V. Räisänen and Arto Lehto

Radio Propagation in Cellular Networks, Nathan Blaunstein

Radio Resource Management for Wireless Networks, Jens Zander and Seong-Lyun Kim

Radiowave Propagation and Antennas for Personal Communications, Third Edition, Kazimierz Siwiak and Yasaman Bahreini

RDS: The Radio Data System, Dietmar Kopitz and Bev Marks

Resource Allocation in Hierarchical Cellular Systems, Lauro Ortigoza-Guerrero and A. Hamid Aghvami

RF and Baseband Techniques for Software-Defined Radio, Peter B. Kenington

RF and Microwave Circuit Design for Wireless Communications, Lawrence E. Larson, editor

Sample Rate Conversion in Software Configurable Radios, Tim Hentschel

Signal Processing Applications in CDMA Communications, Hui Liu

Smart Antenna Engineering, Ahmed El Zooghby

Software Defined Radio for 3G, Paul Burns

Spread Spectrum CDMA Systems for Wireless Communications, Savo G. Glisic and Branka Vucetic

Technologies and Systems for Access and Transport Networks, Jan A. Audestad

Third-Generation and Wideband HF Radio Communications, Eric E. Johnson, Eric Koski, William N. Furman, Mark Jorgenson, and John Nieto

Third Generation Wireless Systems, Volume 1: Post-Shannon Signal Architectures, George M. Calhoun

Traffic Analysis and Design of Wireless IP Networks, Toni Janevski

Transmission Systems Design Handbook for Wireless Networks, Harvey Lehpamer

UMTS and Mobile Computing, Alexander Joseph Huber and Josef Franz Huber

Understanding Cellular Radio, William Webb

Understanding Digital PCS: The TDMA Standard, Cameron Kelly Coursey

Understanding GPS: Principles and Applications, Second Edition, Elliott D. Kaplan and Christopher J. Hegarty, editors

Understanding WAP: Wireless Applications, Devices, and Services, Marcel van der Heijden and Marcus Taylor, editors

Universal Wireless Personal Communications, Ramjee Prasad

WCDMA: Towards IP Mobility and Mobile Internet, Tero Ojanperä and Ramjee Prasad, editors

Wireless Communications in Developing Countries: Cellular and Satellite Systems, Rachael E. Schwartz

Wireless Communications Evolution to 3G and Beyond, Saad Z. Asif

Wireless Intelligent Networking, Gerry Christensen, Paul G. Florack, and Robert Duncan

Wireless LAN Standards and Applications, Asunción Santamaría and Francisco J. López-Hernández, editors

Wireless Sensor and Ad Hoc Networks Under Diversified Network Scenarios, Subir Kumar Sarkar

Wireless Technician's Handbook, Second Edition, Andrew Miceli

For further information on these and other Artech House titles, including previously considered out-of-print books now available through our In-Print-Forever® (IPF®) program, contact:

Artech House
685 Canton Street
Norwood, MA 02062
Phone: 781-769-9750
Fax: 781-769-6334
e-mail: artech@artechhouse.com

Artech House
16 Sussex Street
London SW1V 4RW UK
Phone: +44 (0)20 7596-8750
Fax: +44 (0)20 7630-0166
e-mail: artech-uk@artechhouse.com

Find us on the World Wide Web at: www.artechhouse.com